Signals and Communication Technology

More information about this series at http://www.springer.com/series/4748

Rainer Strobel

Channel Modeling and Physical Layer Optimization in Copper Line Networks

Rainer Strobel
Intel Connected Home Division
Munich, Bayern
Germany

ISSN 1860-4862 ISSN 1860-4870 (electronic)
Signals and Communication Technology
ISBN 978-3-030-06254-5 ISBN 978-3-319-91560-9 (eBook)
https://doi.org/10.1007/978-3-319-91560-9

Softcover re-print of the Hardcover 1st edition 2019

Printed on acid-free paper

This Springer imprint is published by the registered company Springer International Publishing AG part of Springer Nature
The registered company address is: Gewerbestrasse 11, 6330 Cham, Switzerland

Preface

The twisted pair access network is still the main technology for fixed Internet access, and even most of the wireless data traffic is offloaded via Wi-Fi and other wireless home networking technologies to the fixed access network. To satisfy the increasing demand for high bandwidth in the access network while keeping the deployment costs low, hybrid networks of fiber to the distribution point (FTTdp) are a promising technology and give an interesting field of innovation.

To achieve gigabit per second data rates on low-quality phone wires, significant improvements of the DSL technology are required. Besides higher data rates, low power consumption is another important requirement to enable the design of small distribution point units without local power supply. The G.fast technology is designed to address both, higher data rates and low power consumption.

This book covers three major fields of innovation, which are the basis for the G.fast technology: the characteristics of the twisted pair copper channel and MIMO channel modeling at the frequencies used for G.fast, physical layer optimization to achieve high data rates on twisted pair copper cable binders, and discontinuous operation to reduce power consumption.

The work on this dissertation started in 2011 at the time, when discussions in the ITU standardization on a next-generation DSL technology after VDSL2 Vectoring started. This gave the opportunity to work on the technology from the first pre-standardization investigations toward the availability of first products implementing the G.fast technology. It was my intention to cover implementation as well as research aspects in this work and to bring recent signal processing and communications research into practice.

I would like to thank Professor Wolfgang Utschick and the colleagues at the university for good discussions, ideas, and an inspiring environment. At the company, Intel, I would like to thank my colleagues for their support which allowed me to get insights into the "real world" of chip development and systems engineering for access networks. And I would like to thank my family and my friends for their support.

Munich, Germany
March 2018

Rainer Strobel

Contents

Acronyms

ADC	Analog-to-digital converter
ADSL	Asynchronous digital subscriber line
AFE	Analog front-end
ATIS	Alliance for Telecommunications Industry Solutions
AWGN	Additive white Gaussian noise
BER	Bit error rate
BS	Bit swap
BT	British Telecom
CA	Crosstalk avoidance
CDF	Cumulative density function
CE	Cyclic extension
CP	Cyclic prefix
CS	Cyclic suffix
CU	Coefficient update
DAC	Digital-to-analog converter
DC	Direct current
DF	Data frame
DFE	Digital front-end
DFT	Discrete Fourier transform
DMT	Discrete multi-tone
DO	Discontinuous operation
DOI	Discontinuous operation interval
DP	Distribution point
DPBO	Downstream power back-off
DPC	Dirty paper coding
DPU	Distribution point unit
DRA	Dynamic resource allocation
DSL	Digital subscriber line
DTAG	Deutsche Telekom
DTU	Data transmission unit

EL-FEXT	Equal length far-end cross talk
ETSI	European Telecommunication Standards Institute
FDD	Frequency-division duplexing
FEXT	Far-end cross talk
FFT	Fast Fourier transform
FRA	Fast rate adaptation
FTTB	Fiber to the building
FTTC	Fiber to the curb
FTTdp	Fiber to the distribution point
FTTH	Fiber to the home
FTU	Fast transceiver unit
FTU-O	FTU at the optical network side
FTU-R	FTU at the remote side
GPON	Gigabit passive optical network
ISO	International Organization for Standardization
ITU	International Telecommunication Union
LD	Line driver
LMS	Least mean squares
LSB	Least significant bit
MIMO	Multiple input–multiple output
MISO	Multiple input–single output
MMSE	Minimum mean squared error
MSB	Most significant bit
MSE	Mean squared error
NEXT	Near-end cross talk
NOI	Normal operation interval
OSI	Open Systems Interconnection
PAM	Pulse amplitude modulation
PBO	Power back-off
PE	Polyethylene
PMD	Physical media dependent
PMS-TC	Physical media specific transmission convergence
PON	Passive optical network
PSD	Power spectral density
QAM	Quadrature amplitude modulation
RMC	Robust management channel
RPA	RMC parameter adjust
RS	Reed–Solomon code
SDSL	Synchronous digital subscriber line
SF	Sync frame
SHDSL	Synchronous high-speed digital subscriber line
SISO	Single input–single output
SNR	Signal-to-noise ratio
SRA	Seamless rate adaptation
SS	Sync symbol

SU	Signal update
TCM	Trellis coded modulation
TDD	Time-division duplexing
TDMA	Time-division multiple access
THP	Tomlinson–Harashima precoding
TIGA	Transmitter-initiated gain adjustment
TOT	Tone ordering table
TPS-TC	Transport protocol specific transmission convergence
UPBO	Upstream power back-off
VDSL	Very high-speed digital subscriber line
VDSL2	Very high-speed digital subscriber line 2
VF	Velocity factor
W-MMSE	Weighted minimum mean squared error

Chapter 1
Introduction to Fast Digital Access and FTTdp Networks

Broadband Internet access is an integral part of modern life. More and more private activities and work tasks depend on Internet services. This causes the demand for higher data rates in the Internet access network. While wireless broadband experiences an immense growth within the last years, the main portion of Internet data traffic is still transported over the fixed access network. Even most of the wireless data traffic is transported to the subscribers via the fixed Internet access network and locally distributed by a wireless local area network.

To satisfy the increasing demands, the fixed broadband access network is continuously improved by introduction of more advanced data transmission technologies and by changing the network architecture to bring high speed fiber connections closer to the subscribers. The three main transmission media used for fixed broadband access are twisted pair copper wires if the telephony network, coaxial cables and fiber connections of the gigabit passive optical network (GPON).

Copper-based access is by far the most widely used broadband technology among them because of the wide availability of the fixed telephony network [1]. Due to the long evolution of the copper-based access network, there is a variety of coexisting technologies. Figure 1.1 gives an overview on the mix of technologies and network topologies.

The first generation of digital subscriber line technologies is based on ADSL (asynchronous digital subscriber line [2]) and SDSL (synchronous digital subscriber line [3]) transmission technologies. They achieve data rates in the range of 10s of Mbits/s and support a twisted pair line length of several kilometers. ADSL and SDSL are served from a central office, a network node that supplies several thousand subscribers.

To improve data rates and service quality, new network nodes, the street cabinets, and a new transmission technology, VDSL2 (very high speed digital subscriber line 2 [4]), has been introduced. The VDSL2 service achieves 100 Mbit/s data rate and

R. Strobel, *Channel Modeling and Physical Layer Optimization in Copper Line Networks*, Signals and Communication Technology,
https://doi.org/10.1007/978-3-319-91560-9_1

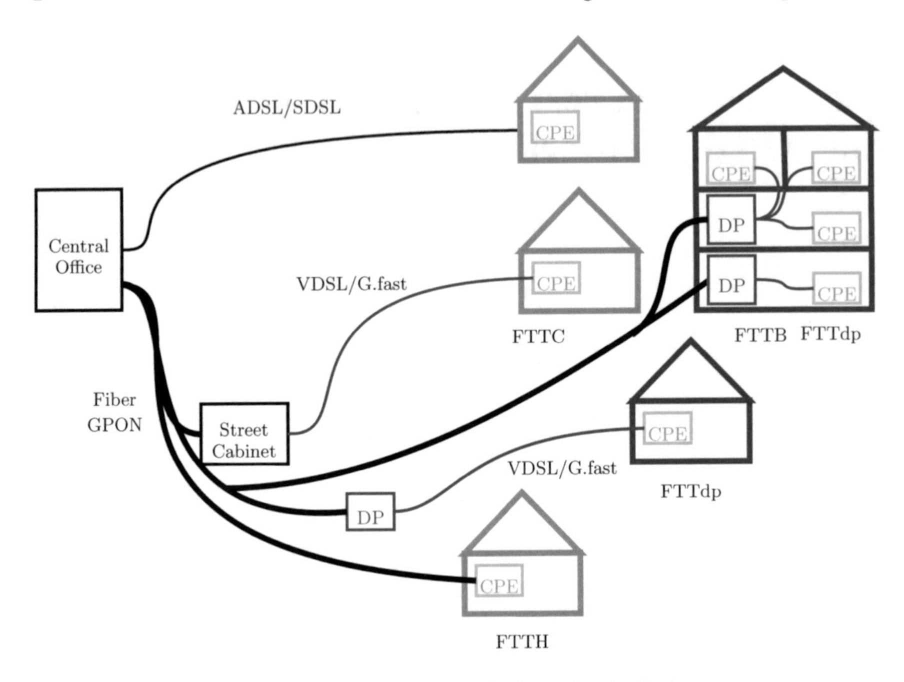

Fig. 1.1 Overview over copper access network topologies and technologies

works on loops shorter than 1000 m. A fiber connection is used to connect the street cabinets with the central office. The street cabinets serve several hundred subscribers. The network topology is called fiber to the curb (FTTC). With the introduction of crosstalk cancelation for VDSL2, 100 Mbit/s DSL services achieve much longer reach.

The next generation network topology is called fiber to the distribution point (FTTdp). The distribution point (DP) is a network node, which has typically a distance below 100 m to the subscriber. At the DP, cable bundles contain a small number of twisted pairs, connecting e.g., 16 or 24 subscribers. The fiber runs to the DP, where a distribution point unit (DPU) is placed. It is supplied with power over the twisted pairs from the subscriber side, as there is typically no local power supply at the distribution point.

Sometimes, FTTdp networks are named fiber to the building (FTTB) in case that the distribution point is inside a building or even fiber to the home (FTTH) when the distribution point serves only a single household.

The transmission technology to be used to transport the data from the fiber link over the remaining distance of twisted pair copper wires is G.fast, as standardized by the ITU [5, 6]. Target data rates are 1 Gbit/s or even 2 Gbit/s on very short loops [7].

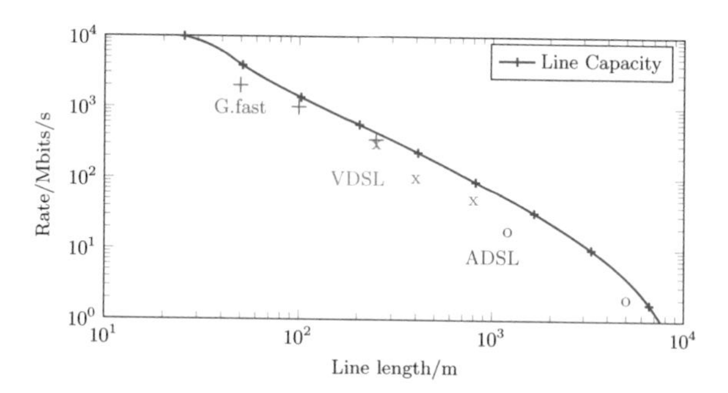

Fig. 1.2 The capacity of a single twisted pair as a function of line length

The general tendency of copper access technologies is shown Fig. 1.2.[1] The required data rates increase continuously, while the copper access network follows this demand by shortening the length of the copper wires between optical network unit and subscriber. More recent technologies allow higher rates at shorter line length by using a wider spectrum. Past technologies like ADSL or VDSL as well as the current G.fast systems are optimized to achieve optimal performance at their target line length, which is for G.fast around 100 m achieving up to 400 m.

This work focuses on the FTTdp network and the corresponding physical layer technology, G.fast. The key requirements of G.fast FTTdp are

- deployment flexibility,
- low power consumption,
- high data rates and low latency.

The flexibility of deployment is achieved by the reverse power feeding architecture where no local power supply is needed at the DPU. The DPU shall be very small, does not require active cooling and can be placed in a water-proof outdoor housing. This leads to the requirement of low power consumption to allow passive cooling and reverse power feeding.

Finally, the technology must be capable to transport high data rates at low latency to give a fiber-like broadband experience while keeping the advantages of a copper technology like customer self install and high quality of service.

1.1 Transmission Technologies

The technology of choice for the FTTdp network is standardized in the ITU-T G.9700 [6] and ITU-T G.9701 [5] standards also known as G.fast.Besides G.fast, there are various legacy technologies deployed. A new technology introduced in the access network must prove to be compatible with the legacy technologies.

---

[1] Capacity is calculated for a single twisted pair line of cable type CAD55 [5]. The aggregate transmit power is limited to 20 dBm, the background noise is at −140 dBm/Hz and the bandwidth is infinite

Pointing to Fig. 1.1, there are legacy technologies such as VDSL2 [4], ASDL and SDSL deployed. Passive optical networks (PON) or coaxial cable networks may also coexist with the copper network as an alternative Internet access technology.

1.1.1 SDSL

Synchronous digital subscriber line (SDSL, SHDSL [3]) technologies are based on single-carrier baseband modulation. They use pulse amplitude modulation (M-PAM) with different constellation sizes. Synchronous DSL transceivers send upstream and downstream signals synchronously, meaning with the same symbol rate at the same time and frequency (full duplex transmission). The full duplex scheme uses the channel more efficient than time or frequency duplexing, because it is not required to allocate separate frequencies or time for each transmission direction.

At each receiver, the near-end transmit signal is canceled out by echo cancelation. SDSL transceivers introduced forward error correction using trellis coding [8]. Tomlinson precoding [9] for pre-distortion of the channel introduced to compensate the channel distortion. SDSL technology is widely used for business applications, where symmetric services with equal upstream and downstream rates are needed.

Private users usually require high downstream and lower upstream data rates and therefore use an asynchronous DSL service.

1.1.2 ADSL/VDSL

Asynchronous DSL (ADSL) and Very High Speed DSL 2 (VDSL2) use DMT multicarrier modulation. Uplink and downlink signals use different frequencies (frequency division duplexing, FDD). Upstream and downstream frequency bands are fixed in band plans which are part of the corresponding standards (ADSL [2], VDSL2 [4]). There are several profiles available, which allow different upstream/downstream ratios.

VDSL2 supports data rates around 100 Mbit/s using frequencies up to 17.6 MHz. A VDSL2 extension called G.vector [10] defines protocols for crosstalk cancelation. Forward error correction is done with a specific type of trellis code, the 4D Wei code [11], which is combined with an outer Reed-Solomon [12] code.

1.1.3 G.fast

The G.fast standard also uses DMT modulation. Linear crosstalk cancelation is a mandatory part of the G.fast specification. In contrast to VDSL2, G.fast uses time division duplexing (TDD) to separate upstream and downstream transmission. TDD requires the synchronization of all lines in a binder such that upstream and downstream time slots are aligned. But the synchronization is anyway required for crosstalk cancelation.

In terms of power consumption, TDD has advantages, because the transmitter can be switched off during the receive phase and vice versa. Low power consumption is very important for G.fast, because it is designed to be used in a small form-factor DPU allowing power supply by reverse power feeding.

1.1.4 Coexistence

The different access technologies coexist with each other. This is mainly implemented by a spectral separation of the services. VDSL uses a wider spectrum than ADSL and G.fast a wider spectrum than VDSL. In that sense, it is possible to exclude the legacy spectrum from the used frequency bands.

In case that the frequency bands of two technologies overlap, the legacy service is protected with specific spectral constraints. VDSL supports protection of ADSL with a method called downstream power back-off (DPBO) and upstream power back-off (UPBO). UPBO is also applied in VDSL deployments without crosstalk cancelation to protect the upstream signals of longer loops.

1.2 Reference Model

Communication systems are modeled in a layer model, e.g. the OSI ISO model [13]. Referring to the 7-layer OSI-ISO model, the G.fast standard defines a physical layer (layer 1) point-to-point protocol as well as the interface to the layer 2 (data link layer [13]).

The G.fast physical layer is divided into three sub-layers as shown in Fig. 1.3, the transport protocol specific transmission convergence (TPS-TC) sublayer, the physical media specific transmission convergence sublayer (PMS-TC) and the physical media dependent (PMD) sublayer. There are interfaces between the sub-layers as well as the interface to the higher layer and to the physical media.

The TPS-TC sublayer converts the packet based traffic of the higher layer into data transmission units (DTUs). The defined functionality includes the insertion of the overhead channel (the eoc overhead channel [5]) as well as buffering and traffic management towards the higher layer (Layer 2). The interface towards the higher layer is called γ interface and the interface to the PMS-TC sub-layer is the α interface.

The PMS-TC sublayer performs Reed-Solomon coding on the DTUs. At this point, physical layer retransmission buffers are handled. Each DTU contains redundancy information that is used to check whether the DTU has been received correctly. In case of a decoding error, a retransmission of the corresponding data is requested.

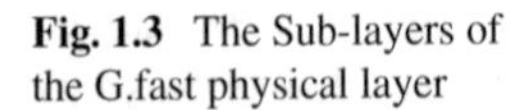

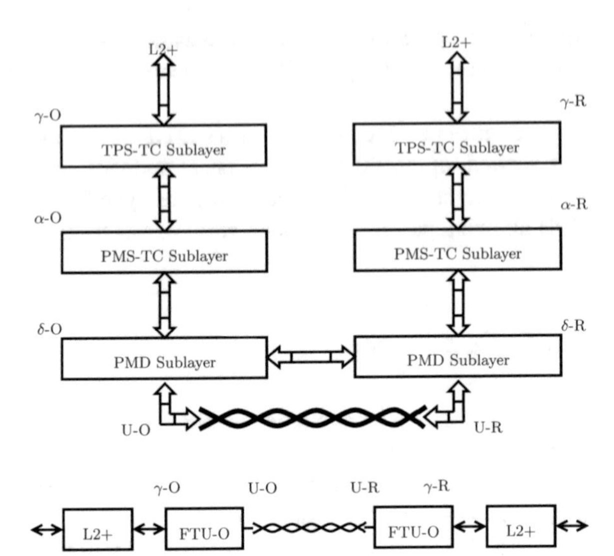

Fig. 1.3 The Sub-layers of the G.fast physical layer

Fig. 1.4 The G.fast reference model according to the G.fast standard [5]

A second overhead channel, the RMC (robust management channel) [5] is inserted at that point. The interface to the PMD layer is the δ-interface.

The PMD sublayer performs trellis coding and includes modulation of the data signals as well as precoding and equalization. The PMD sublayer also performs transmit spectrum shaping and transmitter/receiver optimization tasks. When referring to physical layer optimization, most of these tasks are performed in the PMD sublayer. Transformation between time domain signals and frequency domain signals is also applied at that point, including time domain filtering and analog-to-digital conversion. The interface to the physical medium is the U-interface.

Besides these sub-layers, the G.fast standard defines two network components, referred to as FTU (Fast transceiver unit). The FTU at the DPU is called FTU-O (FTU at the optical network side) and the FTU at the customer premises equipment is called FTU-R (remote side). The corresponding reference model according to [5] is shown in Fig. 1.4.

All the interfaces shown in Fig. 1.3 appear at both ends, the FTU-O and the FTU-R. The G.fast standard defines the transmitter functionality, only. The implementation of the CPE (customer premises equipment) transmitter as well as the DPU transmitter are described in the standard, while the receivers are vendor discretionary.

1.3 G.fast FTTdp Challenges

The main challenges of the G.fast FTTdp network translate into three areas of research, channel modeling, maximization of data rates and minimization of power consumption.

Chapter 2 introduces a channel model for twisted pair copper cable bundles at high frequencies that is suited to G.fast FTTdp. It allows to predict the behavior of the transmission technology in a multi-user MIMO (multiple-input multiple output) environment, including coexistence with legacy technologies. The modeling approach has been standardized in the Broadband Forum standard TR-285 which describes channel models for the FTTdp network [14].

Chapter 3 describes the details of the G.fast physical layer technology, including modulation and coding. The key to achieve increasing performance targets and stay close to the theoretical capacity on short copper loops is an optimized transmission scheme, especially in the multi-user MIMO environment. Focusing on optimized precoding strategies for the downstream direction and optimized equalization in upstream direction. Methods are shown to include specific implementation knowledge and implementation limitations into the optimization process.The investigation is based on the channel models from Chap. 2, allowing a trade-off between complexity and potential performance improvement for twisted pair cable bundles.

Chapter 4 introduces a power consumption minimization approach based on discontinuous operation of the copper link physical layer transmission.The idea is to scale power consumption with the actual data traffic on each line. However, optimization problems have to be solved in a very short time frame.The power minimization framework introduced here combines the spectrum, precoding and equalization optimization of Chap. 3 with a power consumption optimized transmission time optimization.

References

1. OECD:OECD Broadband statistics (2015). http://oecd.org/sti/ict/broadband
2. ITU-T Rec. G.992.5: Asymmetric digital subscriber line 2 transceivers (ADSL2)- Extended bandwidth ADSL2 (ADSL2plus) (2005)
3. ITU-T Rec. G.991.2: Single-pair high-speed digital subscriber line (SHDSL) transceivers (2003)
4. ITU-T Rec. G.993.2: Very high speed digital subscriber line transceivers 2 (VDSL2) (2006)
5. ITU-T Rec. G.9701: Fast Access to Subscriber Terminals - Physical layer specification. ITU Recommendation (2015)
6. ITU-T Rec. G.9700: Fast Access to Subscriber Terminals (FAST) - Power spectral density specification. ITU Recommendation (2013)
7. Oksman, V., Strobel, R., Wang, X., Wei, D., Verbin, R., Goodson, R., Sorbara, M.: The ITU-T's new G. fast standard brings DSL into the gigabit era. IEEE Commun. Mag. **54**(3), 118–126 (2016)
8. Ungerboeck, G.: Channel coding with multilevel/phase signals. IEEE Trans. Inf. Theory **28**(1), 55–67 (1982)
9. Tomlinson, M.: New automatic equaliser employing modulo arithmetic. Electron. Lett. **7**(5), 138–139 (1971)
10. ITU-T Rec. G.993.5-2010: Self-FEXT CANCELLATION (Vectoring) for Use with VDSL2 Transceivers (2010)
11. Wei, L.F.: Trellis-coded modulation with multidimensional constellations. IEEE Trans. Inf. Theory **33**(4), 483–501 (1987)

12. Reed, I., Solomon, G.: Polynomial codes over certain finite fields. SIAM J. Appl. Math. **8**(2), 300 (1960)
13. ITU-T X.200: Information technology - Open Systems Interconnection - Basic Reference Model: The basic model (1994)
14. Broadband Forum TR-285: Cable Models for Physical Layer Testing of G.fast Access Network. Technical report (2015)

Chapter 2
Channel Models for Twisted Pair Cable Bundles

The distribution point network topology in combination with the use of higher frequencies up to 212 MHz in G.fast requires to revisit channel modeling for twisted pair cables. Access cables of the telephone network are not built for such high frequencies, and measurements of access cables in this frequency range show a significant change of the cable characteristics compared to lower frequencies as they are used for ADSL [1] and VDSL2 [2].

With increasing frequency, the simplified modeling approach which was used to design VDSL2 broadband access, e.g., [3] or [4], can no longer be used and a more accurate characterization is required.

Section 2.1 gives an overview over available single line and crosstalk models for twisted pair cable bundles, used for VDSL2 and G.fast system design. The models are compared with measurement data from different sources.

Sections 2.2 and 2.3 present a MIMO modeling approach for twisted pair cable bundles, based on multiconductor transmission line theory. The model is presented in [5] and it is part of the Broadband Forum standard for FTTdp channel modeling, TR-285 [6].

Section 2.4 presents more details of the methods used to match the cable models with measurement data of a specific cable type.

Notation

The following notation is used throughout this chapter. Upper case bold letters denote matrices, e.g., the chain matrix $\boldsymbol{A}$ and the scattering matrix $\boldsymbol{S}$. The operator denotes the Fourier transform. Re(.) gives the real part of a complex number while Im(.) gives the imaginary part of a complex number. $\angle A$ gives the argument of a complex number A and $\mathcal{U}(a, b)$ denotes the uniform random distribution between a and b. $\boldsymbol{I}_L$ is the $L \times L$ identity matrix and $[\boldsymbol{A}]_{ik}$ gives the element in the ith row and kth column of the matrix $\boldsymbol{A}$.

R. Strobel, *Channel Modeling and Physical Layer Optimization in Copper Line Networks*, Signals and Communication Technology,
https://doi.org/10.1007/978-3-319-91560-9_2

2.1 Fundamentals of Twisted Pair Modeling

Mathematical models of signal transmission on transmission lines date back to the Telegraphers equations, developed end of the 19th century [7]. Since that time, twisted pair cable models have been improved continuously, especially due to the increasing bandwidth used for data transmission on twisted pair cables.

This section gives an overview on the channel models developed for VDSL2, which are the basis of G.fast channel models. Besides available VDSL2 channel models, mainly covering frequencies up to 30 MHz, the work is supported by measurement data of twisted pair cable bundles up to 300 MHz which covers the G.fast frequency spectrum.

2.1.1 *Differential Single Line Models*

Channel models for evaluation of data transmission on twisted pair cables are mainly based on a characterization of the differential mode of a single twisted pair. The models describe the primary line parameters, serial resistance R', serial inductance L', parallel capacitance C' and parallel conductance G' per unit length.

From the primary line parameters, an equivalent circuit for a differential length element $\mathrm{d}z$ of the twisted pair line can be described by the circuit shown in Fig. 2.1. While the initial transmission line theory in [7] assumes the primary line parameters to be constant, the channel models for ADSL and VDSL2 require frequency dependent primary line parameters $R'(f)$, $L'(f)$, $C'(f)$ and $G'(f)$ to describe cable characteristics for a higher bandwidth.

Popular models for access cables are the ETSI BT0 model [3] used for VDSL2 up to 30 MHz or the recently introduced ITU model [8, 9] for the approximation of differential mode transfer functions up to 300 MHz. These models describe the primary line parameters by nonlinear functions of frequency f or correspondingly, of the angular frequency $\omega = 2\pi f$.

Following [10], the secondary line parameters, line impedance $Z_0(\omega)$ and propagation term $\gamma(\omega)$ are obtained from the primary line parameters by

$$Z_0(\omega) = \sqrt{\frac{R'(f) + j\omega L'(f)}{G'(f) + j\omega C'(f)}} \tag{2.1}$$

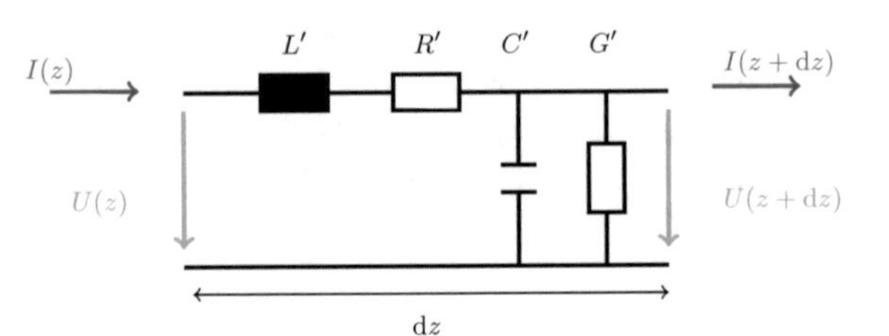

Fig. 2.1 Model of a differential line element $\mathrm{d}z$

and

$$\gamma(\omega) = \sqrt{(R'(f) + j\omega L'(f))(G'(f) + j\omega C'(f))}. \tag{2.2}$$

With that, the voltage $U(z)$ and the current $I(z)$ can be described as a function of the line length z. These are the Telegraphers equations [11] as given by

$$\begin{bmatrix} U(0) \\ I(0) \end{bmatrix} = \begin{bmatrix} \cosh(\gamma z) & Z_0 \sinh(\gamma z) \\ \frac{1}{Z_0} \sinh(\gamma z) & \cosh(\gamma z) \end{bmatrix} \begin{bmatrix} U(z) \\ I(z) \end{bmatrix}, \tag{2.3}$$

which describe the voltage U and current I at the line input as a function of voltage and current at the line output. The matrix description in Eq. (2.3) is referred to as chain matrix description with the chain matrix $\mathbf{A}$ as described in Appendix A.7.

The primary line parameters as shown in Fig. 2.1 form a complex series impedance $Z_s(f) = R'(f) + j\omega L'(f)$ and a parallel admittance $Y_p(f) = G'(f) + j\omega C'(f)$, which are described by the parametric line models. One of them is the ETSI BT0 model [3]. It has been developed for VDSL2 cable modeling and describes the primary line parameters by functions for the serial impedance $Z_s(f)$

$$Z_s(f) = \sqrt[4]{R_{0c}^4 + a_c f^2} + j\omega \left(\frac{L_0 + L_\infty (f/f_m)^{N_b}}{1 + (f/f_m)^{N_b}} \right) \tag{2.4}$$

and for the parallel admittance $Y_p(f)$

$$Y_p(f) = \left(g_0 f^{N_{ge}}\right) + j\omega \left(C_\infty + \frac{C_0}{f^{N_{ce}}} \right). \tag{2.5}$$

Equations (2.4) and (2.5) describe the frequency dependent cable characteristics by eleven constants R_{0c}, a_c, L_0, L_∞, f_m, N_b, g_0, N_{ge}, C_∞, C_0 and N_{ce}. These parameters are selected such that the model results match a specific measured cable type.

It must be noted that the components of the serial impedance $Z_s(f)$ in Eq. (2.4), $R'(f)$, inductance $L'(f)$ are independently modeled from each other. The same holds for $G'(f)$ and capacitance $C'(f)$ in the parallel admittance $Y_p(f)$. This is a drawback when using the cable model for simulations in time domain, because the resulting impulse response may not be causal, depending on the selected parameters. The causality issue is investigated in [12].

A more recent single line model is referred to as ITU model [8] as it is part of the ITU G.fast standard [9]. The ITU model guarantees causal impulse responses, is verified to give a good approximation of channel conditions up to 300 MHz, and covers the full G.fast spectrum.

It uses the following approximation functions for $Z_s(\omega)$

$$Z_s(\omega) = j\omega L_{s\infty} + R_{s0} \Bigg(1 - q_s q_x + \sqrt{q_s^2 q_x^2 + 2\frac{j\omega}{\omega_s} \left(\frac{q_s^2 + j\omega/\omega_s q_y}{q_s^2/q_x + j\omega/\omega_s q_y} \right)}\Bigg) \tag{2.6}$$

and for $Y_p(f)$

$$Y_{\mathrm{p}}(\omega) = j\omega C_{p0}(1 - q_c)\left(1 + \frac{j\omega}{\omega_d}\right)^{-2\phi/\pi} + j\omega C_{p0} q_c. \tag{2.7}$$

The following additional functions are defined for the model:

$$L_{s\infty} = \frac{1}{\eta_{\mathrm{VF}} c_0} Z_{0\infty}\, ,\, C_{p0} = \frac{1}{\eta_{\mathrm{VF}} c_0} \frac{1}{Z_{0\infty}}, \tag{2.8}$$

$$q_s = \frac{1}{q_H^2 q_L}\, ,\, \omega_s = q_H^2 \left(\frac{4\pi R_{s0}}{\mu_0}\right). \tag{2.9}$$

The corresponding parameters to describe the model are R_{s0}, η_{VF}, $Z_{0\infty}$, q_H, q_L, q_x, q_y, ϕ q_c and ω_d and μ_0 is the permeability of free space. Again, the values of these parameters are derived from measurement data. Some of the parameters can be measured directly, e.g., the resistance R_{s0} at $f = 0$ or the line impedance $Z_{0\infty}$ at $f \rightarrow \infty$.[1] Other parameters are derived from a measurement of the cable primary or secondary line parameters by a curve fitting approach.

Besides the two described single line models from [3, 9], which are described in more detail in Appendix A.1, there are many other models of twisted pair cables, which are not discussed, here.

While these single line models are important for performance evaluation, they do not describe the fact that in most cases, twisted pair lines are part of a cable bundle of multiple twisted pairs, resulting in a MIMO channel. The electromagnetic coupling between individual lines is referred to as crosstalk. The mutual couplings between the twisted pairs do not only cause crosstalk, they also affect the behavior of the direct transmission channel.

2.1.2 Crosstalk Models

When signals are transmitted on one pair of a twisted pair cable bundle, the signal partially couples into the other twisted pairs of the binder. This phenomenon is referred to as crosstalk. We distinguish between near-end crosstalk (NEXT) and far-end crosstalk (FEXT) as shown in Fig. 2.2. Near-end crosstalk is the coupling between two twisted pairs at the same end of the cable bundle, while FEXT is the coupling between two twisted pairs at the opposite ends of the cable bundle.

The initial approach of crosstalk modeling in DSL systems described crosstalk as an additional noise source at the receiver. The crosstalk signal level observed on a certain victim line v depends on the coupling between the disturber line d and the victim line and on the signal transmitted by the disturber line. In this context,

[1] The line impedance converges to the value of $Z_{0\infty}$ within the measurement precision at the measured frequencies, as explained in Sect. 2.4.3.

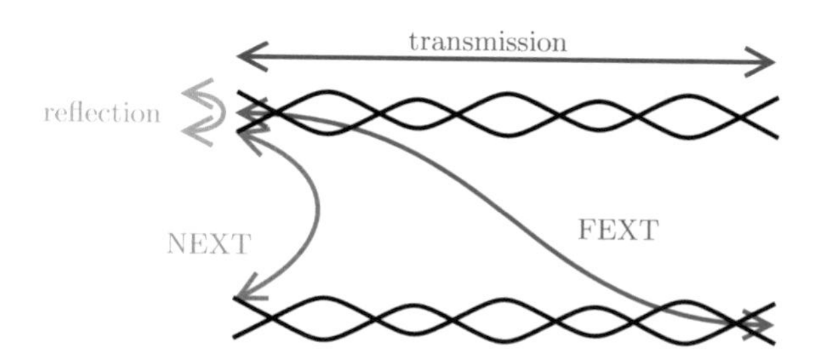

Fig. 2.2 Couplings between two twisted pair wires

the term self-crosstalk describes crosstalk between multiple lines using the same transmission methods and therefore the same type of signals. Alien crosstalk is the crosstalk between different transmission technologies, e.g., between VDSL2 and G.fast systems.

In [3], VDSL2 self crosstalk is modeled as noise with a certain noise spectrum that depends on the cable characteristics and the number of disturbing lines. Crosstalk is a statistical phenomenon. Widely used definitions e.g., in [13] or [3], use a worst case model, describing the crosstalk power that is not exceeded with 99% probability. The 99% worst-case crosstalk coupling $|H_{vd,99}(f,z)|^2$ according to [3] is given by

$$|H_{vd,99}(f,z)|^2 = |H_v(f,z)|^2 \cdot \kappa^2 \cdot z_{\text{coupling}} \cdot f^2 \tag{2.10}$$

where $H_v(f,z)$ is the direct channel transfer function of the victim line v, z_{coupling} is the coupling length between the lines, and κ is a scaling factor (see Appendix A.2).

In case of multiple disturbers, the crosstalk spectrum is summed up over the individual disturbers [14]. Assuming L disturbers with transmit power-spectral densities (PSD) $\psi_1(f)$ to $\psi_L(f)$, the sum noise PSD $\psi_{\text{sum}}(f)$ according to [14] is given by

$$\psi_{\text{sum}}(f) = \left(\sum_{l=1}^{L} \left(|H_{vd,99}(f,z)|^2 \psi_l(f) \right)^{K_n} \right)^{\frac{1}{K_n}}. \tag{2.11}$$

The empirical parameter K_n is defined in [3] to be $K_n = \frac{1}{0.6}$.

A crosstalk model where crosstalk is considered as noise is no longer applicable when MIMO signal processing is performed at the transmitter or receiver of a multi-line DSL system such as a street cabinet or a DPU. For the analysis of crosstalk cancelation for VDSL2 [15], MIMO models have been introduced, describing a crosstalk matrix $\boldsymbol{H}(f)$, which contains the direct transmission channels $H_v(f)$ for each line v and crosstalk couplings $H_{vd}(f)$ from each disturber line d to each victim line v.

The ATIS MIMO model [4] is based on the 99% model of Eq. (2.10) and extends that by a frequency independent crosstalk strength matrix $\boldsymbol{X}_{\text{dB}}$. The crosstalk coupling path from disturber line d to victim line v is modeled as

$$H_{vd}(f) = |H_v(f)| f e^{j\varphi(f)} \kappa \sqrt{z_{\text{coupling}}} 10^{x_{\text{dB}\,vd}/20} \tag{2.12}$$

with a coupling length z_{coupling} between the disturber and the victim line and a random phase term $\varphi(f) = \angle H_v(f) + \varphi_0$. $\varphi_0 \sim \mathcal{U}(0, 2\pi)$ is a frequency-independent random phase. The coupling strength matrix $\mathbf{X}_{\text{dB}}$ is created from a statistical analysis of crosstalk measurement as shown in [4].

A similar modeling approach is used in the crosstalk model of TR-285, Annex A [6]. Crosstalk coupling according to [6] is given by

$$H_{\text{FEXT}\,vd}(f) = |H_v(f)| 10^{A_{\text{dB}}(f)/20} e^{j\varphi(f)} \kappa \sqrt{z_{\text{coupling}}} 10^{x_{\text{dB}\,vd}/20} \tag{2.13}$$

with the weighting function $A_{\text{dB}}(f)$ that is used to approximate the frequency dependency of crosstalk at high frequencies. While in Eq. (2.12), $\frac{|H_{vd}(f)|^2}{|H_v(f)|^2} \sim f^2$ holds, under certain conditions, the proportionality $\frac{|H_{vd}(f)|^2}{|H_v(f)|^2} \sim f^4$ has been observed, e.g., in well structured quad cables [16] or high frequencies [17]. To approximate this behavior, $A_{\text{dB}}(f)$ is defined in [6] to be

$$A_{\text{dB}}(f) = \begin{cases} 20 \log_{10}(f) & \text{for } f \leq 75\,\text{MHz} \\ -157.4 + 40 \log_{10}(f) & \text{for } f > 75\,\text{MHz}. \end{cases} \tag{2.14}$$

MIMO channel models such as Eqs. (2.12) and (2.13) describe MIMO twisted pair channels with low complexity. However, there are several drawbacks of this approach, which are more obvious when comparing the model with measurement data of twisted pair cables at high frequencies (see Sect. 2.1.3).

In opposite to the single line models as described in Sect. 2.1.1, the crosstalk couplings of Eqs. (2.12) and (2.13) do not provide an equivalent circuit and the corresponding MIMO description, e.g., in a chain matrix $\boldsymbol{A}(f)$. Only the crosstalk transfer functions are described, but not the near-end crosstalk and reflections. Therefore, it is not possible to cascade channel matrices from the ATIS model.

Another drawback that results from the this modeling approach is that the model is not guaranteed to be passive. The crosstalk transfer functions are "added" to the direct channel single line model, but no interaction between direct channel and crosstalk is considered. The more lines there are in the binder, the more sum power $\sum_{d=1}^{L} |H_{vd}(f)|^2$ is received at a line v. Furthermore, Eqs. (2.12) and (2.13) increase with increasing frequency without limit. Both effects may result in transfer functions where more power is received at the far end than transmitted.

2.1.3 Cable Measurements at High Frequencies

This section presents measurement data of twisted pair cables at G.fast frequencies. Data from different sources is presented. A measurement campaign at Deutsche Telekom [18] is the basis for the channel modeling approach presented in [5, 6], containing measurements of four different cables. Three of them, named DTAG-PE05, DTAG-PE06 and DTAG-YSTY are evaluated in more detail in Appendix A.4.

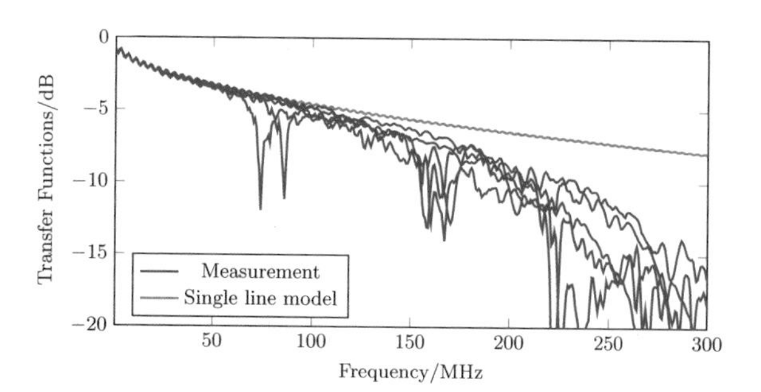

Fig. 2.3 Comparison between measured direct channel transfer functions and single line model transfer functions of DTAG-PE05 cable with 30 m length and 10 pairs

More cable measurement data is available from the Netherlands [19], and Swisscom [20]. These reports contain measurements of direct channel and FEXT transfer functions, only. Therefore, they cannot be used to create a cable model. British Telecom provides FEXT [21, 22] and NEXT [23] data to create cable models for the British access network.

From the research projects [24, 25], measurement data of German cables, of the cable types DTAG-YSTY, DTAG-PE05 and DTAG-PE04 are available. From [26], single-ended[2] measurements of a cable referred to as HSA-YSTY cable are available.

Appendix A.4 shows some representative crosstalk and direct channel transfer functions of the measurements which are used to derive parameters for the presented crosstalk model.

The measurement data shows several effects which are not modeled well with the approach of the ATIS model [4] or the TR-285 Annex A model [6]. The first effect to be pointed out is the interaction between crosstalk and the direct channel. The models described in Sect. 2.1.2 use the same direct channel transfer functions for the single line and the MIMO case. However, strong crosstalk couplings which are close to the direct channel attenuation cause a significant loss of signal on the direct channel due to crosstalk and radiation losses. Figure 2.3 shows this effect on the measured direct channel transfer functions of the DTAG-PE05 cable. At higher frequencies, the direct channel attenuation is much higher than predicted by the single line model due to the mentioned losses. While a single line, which is not part of a cable binder, would behave as predicted by the single line model, the electromagnetic coupling between the lines of a cable binder changes the direct channel attenuation as it is shown by the measured transfer functions in Fig. 2.3.

This effect has been reported in [20]. Besides the crosstalk losses, Fig. 2.3 also indicates that the direct channel attenuation is different for different twisted pairs in the cable binder, which requires a random component of the direct channel of the cable binder model.

[2]More details on single-ended and differential modes of twisted pair cables can be found in Appendix A.9.

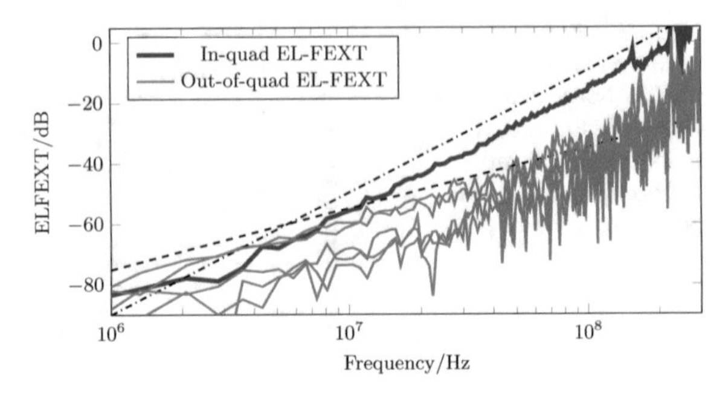

Fig. 2.4 Comparison between measured in-quad and out-of-quad ELFEXT for one line of the DTAG PE05 cable with 10 pairs (5 quads), 30 m length

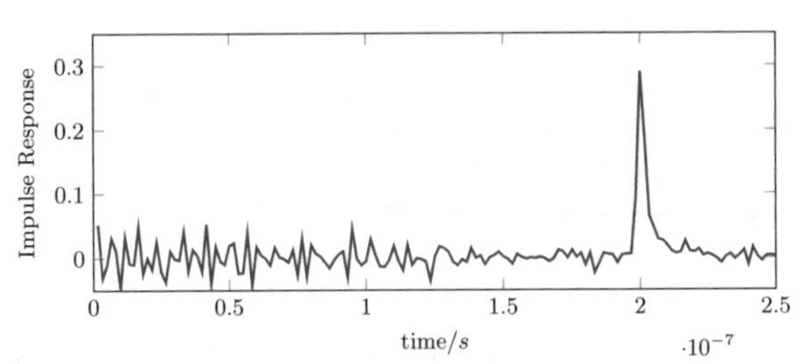

Fig. 2.5 Impulse response of an open-ended PE06 line

For quad-structured cables, the crosstalk behavior between two twisted pairs of a quad shows very different behavior than other crosstalk couplings. This is another effect not covered in the models from Sect. 2.1.2, which can be observed from measurements of the DTAG-PE05 cable, e.g., in Fig. A.1 in Appendix A.4, but also in [20].

The difference between in-quad and out-of-quad crosstalk is illustrated by equal-length FEXT (EL-FEXT) frequency dependency [16]. EL-FEXT is defined to be the crosstalk transfer function divided by the direct channel transfer function, e.g., $\frac{|H_{vd}(f)|^2}{|H_v(f)|^2}$ for the EL-FEXT from line d into line v. For the DTAG-PE05 cable, EL-FEXT is shown in Fig. 2.4. On the logarithmic scale, the different frequency dependencies of in-quad crosstalk (EL-FEXT $\sim f^4$) and out-of-quad crosstalk (EL-FEXT $\sim f^2$) are observed by comparing with the dashed reference line representing the f^2-slope and the dash-dotted line, representing a f^4-slope.

Furthermore, the line parameters, e.g., the line impedance Z_0 are not constant over the length of the twisted pair wire, as Fig. 2.5 indicates. It shows the impulse response of the reflection path of a 20 m long open-ended DTAG-PE06 line. A change of impedance from Z_1 to Z_2 results in a reflection coefficient Γ un-equal zero $\Gamma = \frac{Z_2-Z_1}{Z_2+Z_1}$. Most of the signal is reflected at the open end with $Z \rightarrow \infty$, corresponding to the peak around $2 \cdot 10^{-7}$ µs. Parts of the signal are reflected earlier, indicating an impedance change in the cable.

Based on these observations, the next sections describe a channel model for the FTTdp network, which is based on multiconductor transmission line theory and allows to reproduce the observed effects.

Fig. 2.6 Cable binder segments for two twisted pairs

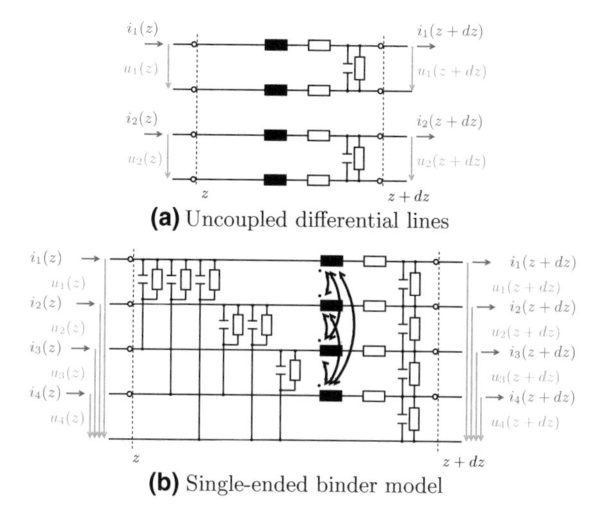

(a) Uncoupled differential lines

(b) Single-ended binder model

2.2 Spatial Domain Transmission Line Model

The proposed channel model which describes the physical characteristics of twisted pair cable bundles is derived from extending the transmission line theory of Sect. 2.1.1 to a multiconductor transmission line. The mathematical background is described in [27] and well investigated in [10].

2.2.1 *Equivalent Circuit*

The extension from the single-line to the multiconductor transmission line theory is demonstrated in Fig. 2.6. Multiple uncoupled twisted pair lines are described by the equivalent circuit of Fig. 2.6a for a differential length element of the line. The circuit of Fig. 2.6a represents the differential transmission mode of the twisted pair lines. Single line models as well as the crosstalk models in Sect. 2.1.2 are based on this circuit.

Extending the circuit to a fully coupled multiconductor network and adding a reference potential gives the circuit of Fig. 2.6b. This is the single-ended[3] description of the cable bundle, where the voltages and currents of individual wires are used and are referred to a common ground potential.

The multi-port chain matrix, an extension of Eq. (2.3), is defined as

$$\begin{bmatrix} \boldsymbol{u}(0) \\ \boldsymbol{i}(0) \end{bmatrix} = \begin{bmatrix} \boldsymbol{A}_{11} & \boldsymbol{A}_{12} \\ \boldsymbol{A}_{21} & \boldsymbol{A}_{22} \end{bmatrix} \cdot \begin{bmatrix} \boldsymbol{u}(z) \\ \boldsymbol{i}(z) \end{bmatrix} \tag{2.15}$$

[3]More details on single-ended and differential mode of twisted pairs can be found in Appendix A.9.

and describes the relation from an input voltage vector $\boldsymbol{u}(0) = [U_1(0), \ldots, U_{2L}(0)]^{\mathrm{T}}$ and an input current vector $\boldsymbol{i}(0) = [I_1(0), \ldots, I_{2L}(0)]^{\mathrm{T}}$ to the output voltage vector $\boldsymbol{u}(z) = [U_1(z), \ldots, U_{2L}(z)]^{\mathrm{T}}$ and output current vector $\boldsymbol{i}(z) = [I_1(z), \ldots, I_{2L}(z)]^{\mathrm{T}}$ for a cable bundle with L twisted pairs, consisting of $2L$ individual wires. The matrix definitions are described in more detail in Annex A.7. The chain matrix $\boldsymbol{A}$ is partitioned into four block matrices $\boldsymbol{A}_{11}$ to $\boldsymbol{A}_{22}$.

Following the transmission line theory for single lines, the chain matrix is derived from the secondary line parameters according to Eq. (2.3). The secondary line parameters are calculated from the primary line parameters using Eqs. (2.1) and (2.2).

According to the multiconductor transmission line theory [10], the primary line parameters are replaced by primary line parameter matrices for resistance $\boldsymbol{R}'(f)$, inductance $\boldsymbol{L}'(f)$, conductance $\boldsymbol{G}'(f)$ and capacitance $\boldsymbol{C}'(f)$ per unit length. For a single-ended description of the cable bundle, the matrices are of size $\boldsymbol{R}'(f), \boldsymbol{L}'(f), \boldsymbol{G}'(f), \boldsymbol{C}'(f) \in \mathbb{R}^{2L\times 2L}$ for a binder with L twisted pairs.

The resistance matrix $\boldsymbol{R}'(f)$ is a diagonal matrix, which depends on the single line resistance $R'(f)$ according to

$$\boldsymbol{R}'(f) = \frac{R'(f)}{2}\boldsymbol{I}_{2L}. \tag{2.16}$$

The factor $\frac{1}{2}$ is required, because the series resistance of the differential single line model is represented by two series resistances, one for each wire, in the single-ended model.

With known permittivity ε and permeability μ of the media between the conductors, the capacitance matrix $\boldsymbol{C}'(f)$ is obtained by matrix inversion [10]

$$\boldsymbol{C}' = \mu\varepsilon\boldsymbol{L}'^{-1}(f) \tag{2.17}$$

from the inductance matrix. Equation (2.17) holds under the assumption of a homogeneous medium between the wires.

With the conductivity σ of the insulation medium, the conductance matrix $\boldsymbol{G}'(f)$ is given by

$$\boldsymbol{G}'(f) = \frac{\sigma}{\varepsilon}\boldsymbol{C}'(f), \tag{2.18}$$

as shown in [10].

According to the multiconductor transmission line theory, the secondary line parameters are derived from the primary line parameters, serial impedance matrix $\boldsymbol{Z}_{\mathrm{s}}(f) = \boldsymbol{R}'(f) + j\omega\boldsymbol{L}'(f)$ and the parallel admittance matrix $\boldsymbol{Y}_{\mathrm{p}}(f) = \boldsymbol{G}'(f) + j\omega\boldsymbol{C}'(f)$, with the help of the following eigendecomposition

$$\boldsymbol{Y}_{\mathrm{p}}(f)\boldsymbol{Z}_{\mathrm{s}}(f) = \boldsymbol{T}_l(f)\gamma^2(f)\boldsymbol{T}_l^{-1}(f). \tag{2.19}$$

The matrix product $\boldsymbol{Y}_{\mathrm{p}}(f)\boldsymbol{Z}_{\mathrm{s}}(f)$ is decomposed into the eigenvector matrix $\boldsymbol{T}_l(f)$ and the propagation term diagonal matrix $\gamma(f)$. This is equivalent to Eqs. (2.1) and

(2.2) for the single line case. The line impedance matrix $\boldsymbol{Z}_0$ is given by

$$\boldsymbol{Z}_0(f) = \boldsymbol{Z}_s(f)\boldsymbol{T}_l(f)\gamma^{-1}(f)\boldsymbol{T}_l^{-1}(f) \tag{2.20}$$

and the corresponding admittance matrix $\boldsymbol{Y}_0(f)$, which is the inverse of the impedance matrix $\boldsymbol{Z}_0(f)$, is given by

$$\boldsymbol{Y}_0(f) = \boldsymbol{T}_l(f)\gamma^{-1}(f)\boldsymbol{T}_l^{-1}(f)\boldsymbol{Y}_p(f). \tag{2.21}$$

The cable bundle is partitioned into different segments of length z_{seg} with constant geometry. Each segment is represented by a chain matrix $\boldsymbol{A}_{\text{seg}}(f)$, which is derived from the secondary line parameters according to

$$\boldsymbol{A}_{\text{seg}} = \begin{bmatrix} \boldsymbol{Z}_0\boldsymbol{T}_l \cosh\left(\gamma z_{\text{seg}}\right)\boldsymbol{T}_l^{-1}\boldsymbol{Y}_0 & \boldsymbol{Z}_0\boldsymbol{T}_l \sinh\left(\gamma z_{\text{seg}}\right)\boldsymbol{T}_l^{-1} \\ \boldsymbol{T}_l \sinh\left(\gamma z_{\text{seg}}\right)\boldsymbol{T}_l^{-1}\boldsymbol{Y}_0 & \boldsymbol{T}_l \cosh\left(\gamma z_{\text{seg}}\right)\boldsymbol{T}_l^{-1} \end{bmatrix}. \tag{2.22}$$

Equation (2.22) describes a section of a cable where the primary line parameters are constant over z_{seg}. Due to twisting of the wires, this is not the case in real cables. Furthermore, measurement data, e.g., in Fig. 2.5 indicates there are random variations of the line parameters over length.

The chain matrix of the complete cable binder $\boldsymbol{A}_{\text{sec}}(f)$ is built from a cascade of many short segments according to

$$\boldsymbol{A}_{\text{sec}}(f) = \prod_i \boldsymbol{A}_{\text{seg},i}(f). \tag{2.23}$$

Within the cable binder section, the cable type and the number of pairs does not change, but there are variations of the line parameters. One or more of such section are used to model the access network according to Sect. 2.3.

With these rules derived from the equivalent circuit in Fig. 2.6, the cable binder can be described for given primary line parameter matrices. The primary line parameter matrices are derived from the cable geometry.

2.2.2 *Reference Geometry*

This section describes a spatial domain geometry approach to derive the primary line parameters, where the inductance matrix $\boldsymbol{L}'(f, z)$ is used as a starting point. It is divided into a length-dependent reference inductance matrix $\boldsymbol{L}_{\text{ref}}(z)$ and the frequency dependent inductance $L'(f)$ such that the inductance matrix element row i and column k of $\boldsymbol{L}'(f, z)$ is given by

$$l_{ik}(f, z) = \begin{cases} l_{\text{ref},ik}(z)L'(f) & \text{for } i = k \\ l_{\text{ref},ik}(z)L'(f \to \infty) & \text{otherwise} \end{cases}. \tag{2.24}$$

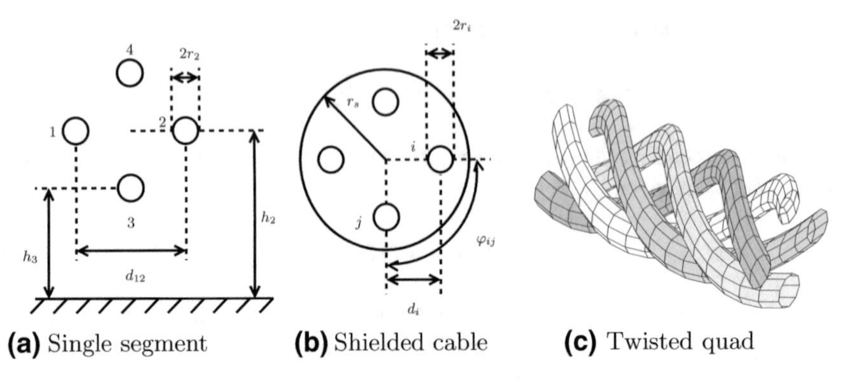

Fig. 2.7 Geometrical model of quad cable [10]

The matrix $\boldsymbol{L}_{\text{ref}}(z)$ represents the cable geometry and the random geometry variations over length, which are frequency-independent, and the single line inductance $L'(f)$ takes care on the frequency dependency of the direct channel which is required to maintain causality of the channel. Only the diagonal elements of $\boldsymbol{L}'(f, z)$ are frequency dependent at the lower frequencies, while the off-diagonal elements are scaled with the inductance value $L'(f \to \infty)$. The single line inductance (Eq. (2.4) or Eq. (2.6)) converges to that value at higher frequencies.[4] This modeling approach is introduced in [5] and part of the cable modeling standard TR-285 Annex B [6].

Reference [28] describes a similar modeling approach for twisted pair cables. Alternatively to [28], where one of the wires is used as reference for the single-ended description, the proposed model uses a separate ground plane as reference for all wires. Reference [28] introduces models of geometry imperfections, while the proposed approach in this work uses a random inductance model.

There are two cable geometries to be taken into account. Unshielded cables with the cross section as shown in Fig. 2.7a, where a separate ground plane is used as a reference potential, and shielded cables as in Fig. 2.7b, where the shield defines the reference potential.

Both geometries can be found in [10] together with the derivation of the inductance matrix with respect to such geometry. The cross sections, as shown in Fig. 2.7a, b are not constant over the length of a twisted pair or twisted quad cable due to the twisting. The twisted quad (see Fig. 2.7c) can be modeled by multiple short slices with constant cross section.

The self inductance l_{ii} of wire i depends on the distance h_i between the wire and the ground plane and on the radius r_i of the wire. The mutual inductance l_{ik} between wires i and k depends on the distance d_{ik} between the wires, in addition. For the unshielded cable geometry Fig. 2.7a, the individual elements of the frequency independent matrix $\boldsymbol{L}_{\text{ref}}(z)$ are given by [10]

[4]For the ITU model, it is one of the derived parameters, $L_{s\infty}$ in Eq. (2.8).

$$l_{\text{ref},ik}(z) = \frac{1}{L'(f \to \infty)} \begin{cases} \frac{\mu}{2\pi} \log\left(\frac{2h_i}{r_i}\right) + l_{\text{rd}} & \text{for } i = k \\ \frac{\mu}{4\pi} \log\left(1 + \frac{4h_i h_k}{d_{ik}^2}\right) + l_{\text{rx}} & \text{for } i \neq k \end{cases}. \tag{2.25}$$

Hereby, $l_{\text{rd}} \sim \mathcal{N}(0, \sigma_{\text{lrd}}^2)$ and $l_{\text{rx}} \sim \mathcal{N}(0, \sigma_{\text{lrx}}^2)$ are random variables to model geometry imperfections. The corresponding random variance σ_{lrd}^2 for the self-inductances and σ_{lrx}^2 for the mutual inductances are derived from measurement data. Examples for some cable types can be found in Appendix A.3.

For shielded cables as in Fig. 2.7b, the following equation from [10] is used

$$l_{\text{ref},ik}(z) = \tag{2.26}$$
$$\frac{1}{L_\infty} \begin{cases} \frac{\mu}{2\pi} \log\left(\frac{r_s^2 + d_i^2}{r_s r_i}\right) + l_{\text{rd}} & \text{for } i = k \\ \frac{\mu}{2\pi} \log\left(\frac{d_k}{r_s} \sqrt{\frac{(d_i d_k)^2 + r_s^4 - 2 d_i d_k r_s^2 \cos\varphi_{ji}}{(d_i d_k)^2 + d_k^4 - 2 d_i d_k^3 \cos\varphi_{ki}}}\right) + l_{\text{rx}} & \text{for } i \neq k \end{cases}$$

where r_s is the radius of the shield, r_i is the radius of the copper wire and φ_{ki} is the angle between the center of wire k and wire i, while d_i is the distance between the center of the cable and the center of wire i.

2.2.3 *Proposed Modeling Steps*

The following modeling steps are performed to generate the chain matrix $\boldsymbol{A}_{\text{seg}}$ of a cable binder segment with the spatial domain MIMO model.

The resistance diagonal matrix $\boldsymbol{R}'(f)$ is derived from the single line model to be

$$\boldsymbol{R}'(f) = \frac{R'(f)}{2} \boldsymbol{I}_{2\text{L}}. \tag{2.27}$$

The inductance matrix $\boldsymbol{L}'(f, z)$ is given by

$$l_{ik}(f, z) = \begin{cases} l_{\text{ref},ik}(z) L'(f) & \text{for } i = k \\ l_{\text{ref},ik}(z) L'(f \to \infty) & \text{otherwise} \end{cases} \tag{2.28}$$

to satisfy the geometry requirements Eq. (2.25), or (2.26) as well as the causality condition. The factor $\frac{1}{2}$ results from the fact that the single line model is a differential model and the present geometry model is a single-ended model.

The reference capacitance matrix $\boldsymbol{C}_{\text{ref}}(z)$ is derived from the reference inductance matrix

$$\boldsymbol{C}_{\text{ref}}(z, f) = \boldsymbol{L}_{\text{ref}}(z)^{-1} \tag{2.29}$$

to satisfy Eq. (2.17). The capacitance matrix $\boldsymbol{C}'(z, f)$ is then

$$\boldsymbol{C}'(z, f) = \boldsymbol{C}_{\text{ref}}(z)C'(f). \tag{2.30}$$

From Eq. (2.18), it follows that the conductance matrix $\boldsymbol{G}'(z, f)$ is given by

$$\boldsymbol{G}'(z, f) = \boldsymbol{C}_{\text{ref}}(z)G'(f). \tag{2.31}$$

Based on the primary line parameter matrices, the chain matrix of a cable binder segment $\boldsymbol{A}_{\text{seg}}$ is calculated, using Eqs. (2.19)–(2.22). An alternative approach, which is applicable when $z_{\text{seg}} \ll \lambda$ holds, is a lumped-circuit model [10]. Then, the segment chain matrix is given by

$$\boldsymbol{A}_{\text{seg}} = \begin{bmatrix} \boldsymbol{I} + z_{\text{seg}}\boldsymbol{Z}_{\text{s}}z_{\text{seg}}\boldsymbol{Y}_{\text{p}} & z_{\text{seg}}\boldsymbol{Z}_{\text{s}} \\ z_{\text{seg}}\boldsymbol{Y}_{\text{p}} & \boldsymbol{I} \end{bmatrix}. \tag{2.32}$$

The cable binder section chain matrix $\boldsymbol{A}_{\text{sec}}$ is derived from the individual segment chain matrices according to Eq. (2.23). In most cases, the transfer functions of the cable binder rather than the chain matrix are required. Therefore, the chain matrix is converted to a scattering matrix according to Eq. (A.18) in Appendix A.8. The model describes the single-ended modes of the cables, but for data transmission, the differential modes are used. The single-ended scattering matrix is converted to the differential mode scattering matrix using Eq. (A.20) in Appendix A.9.

2.3 Network Topology Modeling

The access network connection, as indicated in Fig. 1.1, consists of multiple sections. Telephone cables run from the central office to street cabinets or distribution points and from there into buildings. Within buildings, the cables spread to the individual subscribers, where some in-home wiring section connects to the CPEs. Each section may consist of a different cable type with different characteristics.

The network topology can be represented by three cable sections as shown in Fig. 2.8,

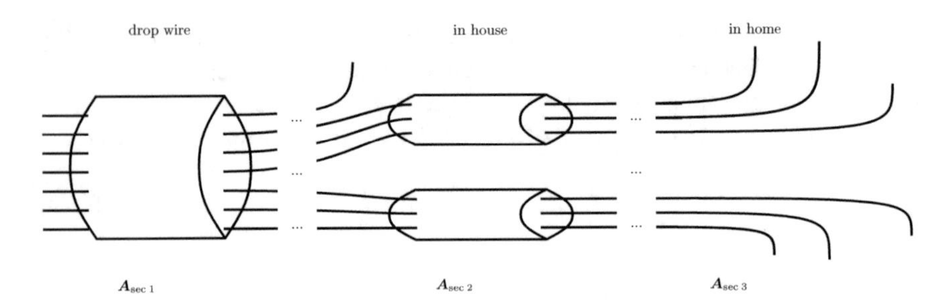

Fig. 2.8 The topology of the telephony network

1. a drop wire section, running from the distribution point to the buildings,
2. an in-building section connecting the drop wire with the individual subscribers homes and
3. an in-home part with a single quad or pair, possibly with bridged taps and similar imperfections.

Each section is described by a matrix $\boldsymbol{A}_{\text{sec},1}$ to $\boldsymbol{A}_{\text{sec},3}$, which is built according to the model of Sect. 2.2. The end-to-end chain matrix $\boldsymbol{A}_{\text{all}}$ is given by

$$\boldsymbol{A}_{\text{all}} = \prod_i \boldsymbol{A}_{\text{sec},i}. \quad (2.33)$$

The conversion from the chain matrix $\boldsymbol{A}_{\text{all}}$ to scattering parameters and the extraction of the differential modes is performed according to Appendix A.8.

2.4 Cable Models and Measurement Data

The single line cable models as well as the crosstalk and MIMO models use a number of cable-specific parameters as input. To describe a certain cable type with the cable model, these parameters are derived from measurement data of the cable.

In a first step, which is presented in Sects. 2.4.1 and 2.4.2, characteristic parameters are derived from the available measurement data. Characteristic parameters are the primary and secondary line parameters for the single line model and crosstalk statistics for the MIMO case.

Some of the model parameters can be measured directly, while other parameters are derived by a regression analysis of the measurement data. This second step is described in Sect. 2.4.3.

2.4.1 Analysis of Single Line Measurement Data

The single line models describe a cable with respect to primary line parameters. But the primary line parameters are not measured directly.

One method to characterize a two-port circuit, such as a single twisted pair line, is the open-short method [18]. For this method, the input impedance of the cable is measured twice. Once with the output shortened, Z_{in}^0, and once with output open, Z_{in}^∞.

With known line length z, the secondary line parameters are derived from the measurement. The line impedance $Z_0(f)$ depends on the open-short measurement according to

$$Z_0^2(f) = Z_{in}^0(f) Z_{in}^\infty(f) \quad (2.34)$$

and the propagation term $\gamma(f)$ is derived using

$$\tanh^2(\gamma(f)z) = \frac{Z_{in}^0(f)}{Z_{in}^\infty(f)}. \tag{2.35}$$

In both cases, the mapping is not unique. Additional knowledge on the line parameters is required. For the impedance, $\text{Re}\{Z_0\} > 0$ and $\text{Im}\{Z_0\} < 0$ holds, which allows a unique mapping for the square root required to derive $Z_0(f)$ from Eq. (2.34). For the propagation term, $\text{Re}\{\gamma\} > 0$ holds and $\text{Im}\{\gamma(f)\}$ is continuously increasing over frequency.

The primary line parameters are then derived according to

$$R'(f) + j\omega L'(f) = Z_s(f) = \sqrt{Z_0^2(f)\gamma^2(f)} \tag{2.36}$$

for the serial impedance $Z_s(f)$ and according to

$$G'(f) + j\omega C'(f) = Y_p(f) = \sqrt{\frac{\gamma^2(f)}{Z_0^2(f)}} \tag{2.37}$$

for the parallel admittance $Y_p(f)$. Again, the mapping is not unique, but each of the primary line parameters is required to be real-valued and positive.

This extraction method has a disadvantage in channels with strong crosstalk, because crosstalk distorts the open-short measurement results. Therefore, it cannot be applied for measurements in the G.fast frequency range and a MIMO extraction method is required to extract the primary line parameters from cable bundles.

2.4.2 *Analysis of MIMO Measurement Data*

Data extraction from MIMO measurements is based on Multiconductor Transmission Line theory [10]. The inverse steps of the channel model described in Sect. 2.2.1 are performed to derive the primary line parameter matrices from measured scattering parameter matrices $\boldsymbol{S}_{\text{meas}}(f)$. The presented idea is investigated in [26].

The method may be performed on the differential mode scattering matrix where the resulting primary line parameters may be used to derive the single line models from Sect. 2.1.1. The single-ended model, as described in Sect. 2.2, is derived from the single-ended scattering matrix.

The measured scattering matrix $\boldsymbol{S}_{\text{meas}}(f)$ is transformed into a chain matrix $\boldsymbol{A}_{\text{meas}}(f)$ by Eq. (A.19) in Appendix A.8

The chain matrix of a cable binder is described in terms of secondary line parameters in Eq. (2.22). The extraction of the diagonal matrix $\boldsymbol{\gamma}(f)$ of transmission terms is based on the eigendecomposition of the matrix $\boldsymbol{A}_{\text{meas},22}$,

$$\boldsymbol{A}_{\text{meas},22} = \boldsymbol{T}_l \boldsymbol{V} \boldsymbol{T}_l^{-1}. \tag{2.38}$$

This is a critical step, as the eigendecomposition is not unique in terms of the order of the eigenvalues, but for the parameter extraction, the secondary and primary line parameters are required to be continuous over frequency. The eigenvalues represent different lines or wires and are re-ordered such that the order of lines does not change over frequency. Equation (2.38) is rewritten as

$$\boldsymbol{A}_{\text{meas},22} = \tilde{\boldsymbol{T}}_l \boldsymbol{\Pi} \boldsymbol{\Pi}^{-1} \tilde{\boldsymbol{V}} \boldsymbol{\Pi} \boldsymbol{\Pi}^{-1} \tilde{\boldsymbol{T}}_l^{-1}, \tag{2.39}$$

which allows an arbitrary permutation matrix $\boldsymbol{\Pi}$ to be inserted. Besides the permutation, the sign may be inverted.

The measurement is performed for multiple frequencies f_k with $k = 1, \ldots, K$ with a sufficiently small frequency steps. With the assumption that $\boldsymbol{T}_l(f_k) \approx \boldsymbol{T}_l(f_{k-1})$ the permutation matrix $\boldsymbol{\Pi}$, which satisfies

$$\min_{\boldsymbol{\Pi}} ||\boldsymbol{T}_l(f_k)\boldsymbol{\Pi} - \boldsymbol{T}_l(f_{k-1})||_2^2 \tag{2.40}$$

is searched. This is a combinatorial optimization over all possible permutation matrices $\boldsymbol{\Pi}$. The optimal permutation gives the final eigenvector matrix $\boldsymbol{T}_l(f_k) = \tilde{\boldsymbol{T}}_l(f_k)\boldsymbol{\Pi}$ and the eigenvalue matrix $\boldsymbol{V}(f_k) = \boldsymbol{\Pi}^{-1}\tilde{\boldsymbol{V}}(f_n)\boldsymbol{\Pi}$.

The diagonal elements of the Eigenvalue matrix $\boldsymbol{V}$ contain the terms $\cosh(\gamma z)$ with the line length z. Therefore, the individual values of the transmission term for each wire i or line i are given by

$$\gamma_i = \frac{\operatorname{acosh}(v_{ii})}{z}. \tag{2.41}$$

With known transmission matrix γ the line admittance matrix $\boldsymbol{Y}_0$ is reconstructed using the $\boldsymbol{A}_{\text{meas},21}$-block matrix of the chain matrix by

$$\boldsymbol{Y}_0 = \boldsymbol{T}_l \sinh(\gamma l)^{-1} \boldsymbol{T}_l^{-1} \boldsymbol{A}_{21} \tag{2.42}$$

and the corresponding impedance matrix is obtained by matrix inversion

$$\boldsymbol{Z}_0 = \boldsymbol{Y}_0^{-1}. \tag{2.43}$$

Equation (2.42) is another critical step due to the inversion of $\sinh(\gamma l)$, which may be close to zero for some frequencies. Frequencies where $\sinh(\gamma l) \approx 0$ may be excluded from the extraction.

From the secondary line parameters, the primary line parameter matrices $\boldsymbol{Z}_\text{s}(f)$ and $\boldsymbol{Y}_\text{p}(f)$ are reconstructed with the help of Eq. (2.19) which gives

$$\boldsymbol{Z}_s = \boldsymbol{R}' + j\omega \boldsymbol{L}' = \boldsymbol{Z}_0 \boldsymbol{T}_l \gamma \boldsymbol{T}_l^{-1} \tag{2.44}$$

for the serial impedance matrix and

$$\boldsymbol{Y}_p = \boldsymbol{G}' + j\omega \boldsymbol{C}' = \boldsymbol{T}_l \gamma \boldsymbol{T}_l^{-1} \boldsymbol{Y}_0 \tag{2.45}$$

for the parallel admittance matrix.

As an example, the result of primary line parameter extraction for the DTAG-YSTY cable is shown in Appendix A.5. The primary line parameter matrices can be used to derive the parameters for single line or MIMO models.

2.4.3 Regression Analysis for Single Line Models

Single line models describe the primary line parameters with nonlinear functions over frequency. To find the appropriate parameters with respect to primary line parameters from measurement data, nonlinear regression tools are applied.

The regression analysis is required, because the measurement data is usually distorted by measurement noise and the relevant parameters, e.g., the primary line parameters, show some randomness due to cable imperfections.

ETSI BT0-Model

The ETSI BT0-model as given by Eqs. (2.4) and (2.5) describes the four primary line parameters $R'(f)$, $G'(f)$, $C'(f)$ and $L'(f)$ independently. Therefore, impulse responses are not guaranteed to be causal, but the parameter fitting by nonlinear regression is more stable due to the small number of parameters involved in each of the four independent optimizations.

The measured primary line parameters are available as a discrete function of frequency, $R'^{(k)}_{\text{meas}}$, $L'^{(k)}_{\text{meas}}$, $C'^{(k)}_{\text{meas}}$ and $G'^{(k)}_{\text{meas}}$ for frequency f_k.

This gives the following four optimization problems to be solved

$$\min_{C_0, C_\infty, N_{ce}} \sum_{k=1}^{K} \left| C'^{(k)}_{\text{meas}} - C_\infty + \frac{C_0}{f_k^{N_{ce}}} \right|^2, \tag{2.46}$$

$$\min_{L_0, L_\infty f_m, N_b} \sum_{k=1}^{K} \left| L'^{(k)}_{\text{meas}} - \frac{L_0 + L_\infty \cdot (f_k/f_m)^{N_b}}{1 + (f_k/f_m)^{N_b}} \right|^2, \tag{2.47}$$

$$\min_{a_c, R_{0c}} \sum_{k=1}^{K} \left| R'^{(k)}_{\text{meas}} - \sqrt[4]{R_{0c}^4 + a_c \cdot f_k^2} \right|^2, \tag{2.48}$$

$$\min_{g_0, N_{ge}} \sum_{k=1}^{K} \left| G'^{(k)}_{\text{meas}} - g_0 \cdot f_k^{N_{ge}} \right|^2 \tag{2.49}$$

for each of the primary line parameters.

The parameter extraction of $R'(f)$ and $G'(f)$ is very sensitive to noise and measurement errors, while the information about the loss of a cable can also be found in the real part of the propagation term $\text{Re}\{\gamma\}$. An alternative solution to derive the loss terms is to solve

$$\min_{a_c, g_0, N_{ge}} \sum_{n=1}^{N} \sum_{l=1}^{L} \left| \mathrm{Re}\{\gamma_{\mathrm{meas}}^{(k)}\} - \mathrm{Re}\{\gamma(f_k)\} \right|^2 \tag{2.50}$$

while the inductance and capacitance parameters are derived in a first step, using Eqs. (2.47) and (2.46) and the DC resistance R_{0c} is determined by a direct resistance measurement. Parameters for the ETSI BT0-model for several cables can be found in Table A.3 in Appendix A.1.

ITU Model

In the ITU model, series resistance and series inductance are described in one equation (see Eq. (2.6)) as well as parallel admittance and parallel capacitance (see Eq. (2.7)). The series impedance $Z_s(f)$ depends on seven parameters η_{VF}, $Z_{0\infty}$, R_{s0}, q_H, q_L, q_x and q_y.

The parallel admittance $Y_p(f)$ depends on five parameters η_{VF}, $Z_{0\infty}$, ϕ, q_c and ω_d. The impedance $Z_{0\infty}$ as well as the velocity factor η_{VF} are parameter to $Z_s(f)$ and $Y_p(f)$. This does not fit well in standard frameworks for nonlinear regression, because the same parameter is part of two different optimizations with different objectives.

It is proposed to derive both values, $Z_{0\infty}$ and η_{VF} from the measurement data. At high frequencies, the value $Z_0(f)$ converges towards $Z_{0\infty}$. Therefore, the parameter $Z_{0\infty}$ is derived according to Sect. 2.4.1 or 2.4.2 as the value of $Z_{0\infty} = Z_0(f \to \infty)$ at the highest measured frequencies.[5] The primary line parameters $L'(f)$ and $C'(f)$ also converge to certain values at high frequencies, e.g., $L_\infty = L'(f \to \infty)$ and $C_\infty = C'(f \to \infty)$. With that, the velocity factor η_{VF} is given by

$$\eta_{\mathrm{VF}} = \frac{1}{c_0 \sqrt{L_\infty C_\infty}}. \tag{2.51}$$

Then, these parameters are not part of the regression analysis.

The optimization problems are

$$\min_{q_H, q_L, q_x, q_y} \sum_{n=1}^{N} \sum_{l=1}^{L} \left| R_{\mathrm{meas}}^{\prime(k)} + j\omega L_{\mathrm{meas}}^{\prime(k)} - Z_s(2\pi f_k) \right|^2, \tag{2.52}$$

$$\min_{q_c, \phi, \omega_d} \sum_{n=1}^{N} \sum_{l=1}^{L} \left| G_{\mathrm{meas}}^{\prime(k)} + j\omega C_{\mathrm{meas}}^{\prime(k)} - Y_p(2\pi f_k) \right|^2 \tag{2.53}$$

with $Y_p(2\pi f_k)$ as defined in Eq. (2.7) and $Z_s(2\pi f_k)$ as in (2.6).

The ITU model parameters extracted for different cable types are summarized in Table A.1 in Appendix A.1.

[5] As indicated, e.g., by Fig. A.6 in Appendix A.5, the impedance converges for typical cables at frequencies below 10 MHz to the value of $Z_{0\infty}$ within the measurement precision.

2.4.4 Multi-line Parameter Fitting

For crosstalk modeling, crosstalk statistics need to be matched between the measured MIMO channel and the channel model. For the crosstalk models described in Sect. 2.1.2, the probability distribution of crosstalk strength X_{dB} is directly used in the models. The crosstalk strength distribution is derived from measured crosstalk, by doing the inverse steps of Eq. (2.12), e.g.,

$$X_{\mathrm{dB},vd} = 10\log_{10}\left(\frac{1}{1+k_{\mathrm{end}}-k_{\mathrm{start}}}\sum_{k=k_{\mathrm{start}}}^{k_{\mathrm{end}}}\left|\frac{H_{vd}^{(k)}}{H_v^{(k)} f \kappa \sqrt{z}}\right|^2\right) \tag{2.54}$$

for a measurement with $k = 1, \ldots, K$ discrete frequency points f_k where averaging is performed from a start frequency index k_{start} to an end frequency index k_{end}. In the model formulation as in Eq. (2.12), the crosstalk strength matrix $\boldsymbol{X}_{\mathrm{dB}}$ is independent of frequency. However, this does not hold true for the whole G.fast frequency range, as already indicated in Eq. (2.13) and it does not hold true for quad-structured cables.

The $\boldsymbol{X}_{\mathrm{dB}}$ matrix evaluated over multiple frequency bands, gives a useful tool to match the crosstalk strength of the model $\boldsymbol{X}_{\mathrm{dB,model}}$ with a certain measured crosstalk strength $\boldsymbol{X}_{\mathrm{dB,meas}}$. The median values $\tilde{X}_{\mathrm{dB,meas}}$ and $\tilde{X}_{\mathrm{dB,model}}$ the crosstalk strength for different frequency points can be used to derive the inductance variance values σ^2_{lrd} and σ^2_{lrx}.

$$\min_{\sigma^2_{\mathrm{lrd}},\sigma^2_{\mathrm{lrx}}} \sum_{k=1}^{K} |\tilde{X}_{\mathrm{dB,meas}} - \tilde{X}_{\mathrm{dB,model}}|. \tag{2.55}$$

One example for the extraction of crosstalk statistics for the DTAG-YSTY cable is shown in Appendix A.6.

The spatial domain crosstalk model from Sect. 2.2 is characterized by inductance matrix variance and the cable geometry. The reference geometry model parameters, which are summarized in Appendix A.3 have been measured on a cable or were taken from the cable specification.

A parameter fit, as for the single line models, is only performed for the random inductance variance values σ^2_{lrx} and σ^2_{lrd}. The resulting parameters are summarized in Appendix A.3 for the spatial domain model.

2.5 Discussion

This chapter presents the work on cable models of twisted pair cables for G.fast FTTdp networks. The channel model is important to evaluate different transmission technology options for G.fast and to evaluate channel capacity and achievable data rates of the FTTdp network.

The available measurement data for twisted pair cables at high frequencies shows characteristics of the channel which are not covered by existing standard models, e.g., in the ETSI model [3] or the ATIS model [4].

The measurement data indicated the following requirements [29] for the MIMO model

- Realistic crosstalk behavior at frequencies beyond 30 MHz, including frequencies where crosstalk is stronger than the direct channel.
- Modeling of the in-quad FEXT of quad-structured cables.
- A random model for the direct channel and the interaction between direct channel and crosstalk.
- The model shall describe a cascade of different cable types, required to build FTTdp network topologies.
- Besides FEXT, NEXT and reflections shall be part of the model to analyze alien crosstalk and impedance change effects.

Multiconductor transmission line theory is the tool to be used for a MIMO cable binder model to satisfy these requirements. Based on the work of [10] on multiconductor transmission line theory and the work of [28], the presented cable model is developed. The focus of this work is to model transmission channels in a realistic way, matching with measurement data of representative access network cables.

The difference in modeling quality is shown in Fig. 2.9, comparing the presented model, which is part of the TR-285 standard as Annex B with the model of TR-285, Annex A, which was developed at the same time, following the approach of the ATIS model [4]. Various contributions were made to develop the channel model in the standard [30–33].

Comparing the dashed lines of the Annex A model in Fig. 2.9 with the solid lines, representing the presented model, the advantages of the geometry-based model are visible when comparing the results with a measurement of the same cable type (10-pair binder of 30 m length, quad-structured DTAG-PE05 cable), as shown in Fig. A.1 in Appendix A.4. The different frequency dependency of in-quad and out-of-quad crosstalk is present in the proposed model and the general frequency dependency matches for both, in-quad and out-of-quad crosstalk with the measurement data. The increased direct channel attenuation due to high crosstalk is present, also and for

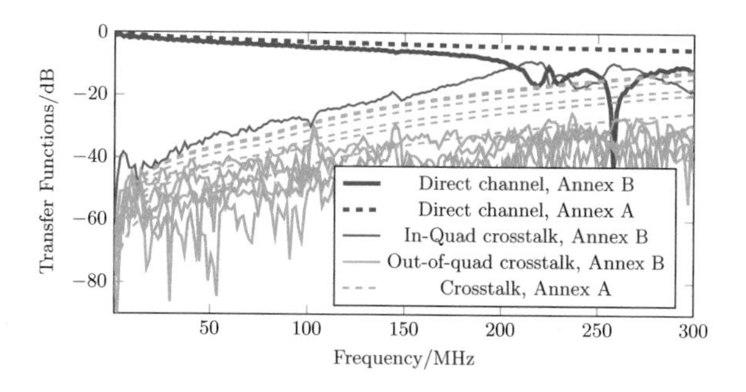

Fig. 2.9 Comparison between the TR-285 Annex A and TR-285 Annex B model for the DTAG-PE05 cable

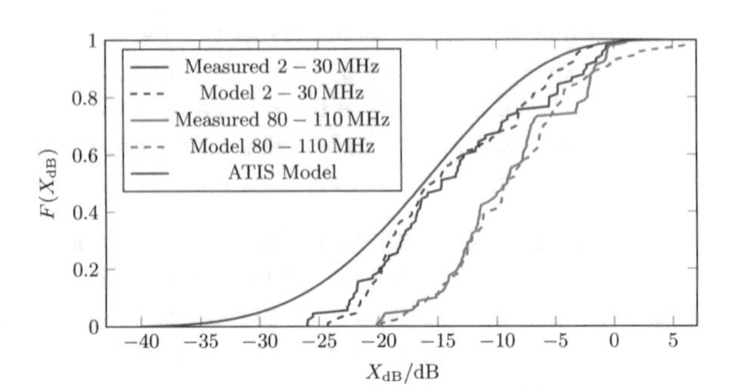

Fig. 2.10 Comparison between crosstalk CDF of measured and modeled DTAG-PE06 cable with 30 m line length and 10 pairs

frequencies above 200 MHz, the behavior of crosstalk that is stronger than the direct channel is present for the proposed model, while this is not the case for the Annex A model.

The crosstalk statistics, represented by the cumulative density function of the elements of the $\boldsymbol{X}_{\mathrm{dB}}$-matrix, is shown in Fig. 2.10 for the DTAG-PE06 cable. At low frequencies, represented by the frequency band between 2 and 30 MHz, there is a good match between the crosstalk distribution of the measurement and the model. For this frequency range, the crosstalk distribution is close to the ATIS model, which is formulated for this frequency range.

When using a higher frequency band between 80 and 110 MHz, the crosstalk is stronger than predicted by the ATIS model, but the spatial domain model still gives a good fit to the measurement data. This shows the capabilities of the spatial domain model to represent crosstalk at high frequencies correctly.

The cable model parameters are available for different types of cables which are representative for the FTTdp network, consisting of in-house cables as well as outdoor cables and high quality and low quality cable types. Line emulation devices, allowing a reproducible MIMO channel based on the presented model for performance testing are in development [34]. The model is publicly available and used for the analysis of different aspects of the G.fast physical layer, e.g., performance of spectrum management and precoding methods in Chap. 3.

References

1. ITU-T Rec. G.992.5: Asymmetric digital subscriber line 2 transceivers (ADSL2)- Extended bandwidth ADSL2 (ADSL2plus) (2005)
2. ITU-T Rec. G.993.2: Very high speed digital subscriber line transceivers 2 (VDSL2) (2006)
3. ETSI TS 101 270-1: Transmission and Multiplexing (TM); Access transmission systems on metallic access cables; Very high speed Digital Subscriber Line (VDSL); Part 1: Functional requirements (2003)
4. Maes, J., Guenach, M., Peeters, M.: Statistical channel model for gain quantification of DSL crosstalk mitigation techniques. In: IEEE International Conference on Communications (ICC) (2009)

5. Strobel, R., Stolle, R., Utschick, W.: Wideband Modeling of Twisted-Pair Cables for MIMO Applications. In: IEEE Global Communications Conference (GLOBECOM) (2013)
6. Broadband Forum TR-285: Cable Models for Physical Layer Testing of G.fast Access Network. Technical report (2015)
7. Heaviside, O.: LXII. On resistance and conductance operators, and their derivatives, inductance and permittance, especially in connexion with electric and magnetic energy. Lond. Edinburgh, and Dublin Philos. Mag. J. Sci. **24**(151), 479–502 (1887)
8. TNO: G.fast: wideband modeling of twisted pair cables as two-ports (2011). ITU-T Contribution SG15/Q4a 11RV-022
9. ITU-T Rec. G.9701: Fast Access to Subscriber Terminals - Physical layer specification (2015). ITU Recommendation
10. Paul, C.: Analysis of Multiconductor Transmission Lines. Wiley-IEEE Press, New York (2007)
11. Starr, T.: DSL Advances. Prentice Hall, Upper Saddle River (2003)
12. Lindqvist, F., Börjesson, P., Ödling, P., Höst, S., Ericson, K., Magesacher, T.: Low-order and causal twisted-pair cable modeling by means of the Hilbert transform. In: AIP Conference Proceedings, vol. 1106, p. 301 (2009)
13. ANSI T1. 417-2003: Spectrum management for loop transmission systems (2003)
14. van den Brink, R.F.M.: Cable reference models for simulating metallic access networks (1998). ETSI STC TM6 Permanent Document TM6(97)02, r970p02r3
15. Ginis, G., Cioffi, J.: Vectored-DMT: a FEXT canceling modulation scheme for coordinating users. In: IEEE International Conference on Communications (ICC), vol. 1, pp. 305–309. IEEE (2001)
16. Muggenthaler, P., Tudziers, C., Berndt, E.: G.fast: EL-FEXT Analysis (2012). ITU-T Contribution SG15/Q4a 2012-06-4A-41
17. van den Brink, R., van den Heuvel, B.: G.fast: dual slope behaviour of EL-FEXT (2012). ITU-T Contribution SG15/Q4a 2012-02-4A-038
18. Kozarev, A., Strobel, R., Leimer, S., Muggenthaler, P.: Modeling of Twisted-Pair Quad Cables for MIMO Applications (2014). Broadband Froum Contribution bbf2014.467
19. TNO: G.fast: far-end crosstak in twisted pair cabling: measurements and modeling (2011). ITU-T Contribution SG15/Q4a 11RV-022
20. Carrick, A., Bongard, T.: G.fast: additional loop sets for simulation purposes (2013). ITU-T Contribution SG15/Q4a 2013-01-Q4-044
21. Humphrey, L., Morsman, T.: G.fast: release of BT cable measurements for use in simulations (2013). ITU-T Contribution SG15/Q4a 2013-1-Q4-066
22. G.fast: Release of BT cable (20 pair) measurements for use in simulations (2015). ITU-T Contribution SG15/Q4a 2015-02-Q4-053
23. Horsley, I., Singleton, H.: Release of BT cable (20 pair) NEXT measurements for use in simulations (2015). ITU-T Contribution SG15/Q4 COM15 - C1499 - E
24. Eder (Lantiq Deutschland GmbH), A., Helfer (CAD Service), G., Leibiger (Fraunhofer ESK), M.: Hochbitratige Access- und Inhausnetze (HAInet) Schlussbericht. Technical report, Forderprogramm IuK Bayern (2014)
25. Maierbacher (Fraunhofer ESK), G., Strobel (Lantiq Deutschland GmbH), R., Liss (InnoRoute GmbH), C.: Entwicklung eines Generators für Mustermodelle verkoppelter Mehrleitersysteme fr die FTTdp-Übertragung. Technical report, Bayerische Forschungsstiftung (2014)
26. Blenk, T.: Entwicklung eines Generators für Mustermodelle verkoppelter Mehrleitersysteme für die FTTdp-Übertragung. Master's thesis, Hochschule Augsburg, University of Applied Sciences (2014)
27. Pipes, L.: X. matrix theory of multiconductor transmission lines. Lond. Edinb. Dublin Philos. Mag. J. Sci. **24**(159), 97–113 (1937)
28. Lee, B., Cioffi, J., Jagannathan, S., Seong, K., Kim, Y., Mohseni, M., Brady, M.: Binder MIMO channels. IEEE Trans. Commun. **55**(8), 1617–1628 (2007)
29. Blenk, T.: Analyse existierender Leitungsmodelle zur Charakterisierung von DSL-Leitungen. Bachelor's thesis, Hochschule Augsburg, University of Applied Sciences (2014)

30. Kozarev, A., Strobel, R.: Update to the Wideband MIMO model of Copper Access Networks up to 300 MHz (2014). Broadband Forum Contribution bbf2014.377.00
31. Leimer, S., Muggenthaler, P., Kozarev, A., Strobel, R.: Some Observations about the Models in SD-285 Section 4.1.5 (2014). Broadband Forum Contribution bbf2014.509.00
32. Kozarev, A., Strobel, R., Leimer, S.: Wideband MIMO model of Copper Access Networks up to 300 MHz (2014). Broadband Forum Contribution bbf2014.117.01
33. Kozarev, A., Strobel, R., Leimer, S., Muggenthaler, P.: Modeling of Twisted-Pair Quad Cables for MIMO Applications (2014). Broadband Forum Contribution bbf2014.467.01
34. Strobel, R.: Line simulator (2014). WO Patent App. WO/2015/132264

Chapter 3
Precoding Optimization for Copper Line Networks

Achieving high data rates on low quality twisted pair cables under certain power and complexity constraints requires optimized physical layer transmission. This chapter discusses signal transmission in G.fast with the focus on precoding and spectrum optimization in downstream direction. A spectrum optimization method for linear [1] and nonlinear zero-forcing [2] precoding as well as a linear MMSE precoder optimization [2] method for G.fast are introduced and compared with other known spectrum optimization methods.

The power constraints to be considered for wireline transmission are different than a sum-power constraint that is mostly used in literature, requiring a more general optimization framework. Furthermore, system-specific limitations such as a limited modulation alphabet are incorporated into the optimization. A realistic estimation of the achievable data rates under the conditions of an implementable G.fast transceiver is provided.

The discussion is based on the FTTdp network structure as shown in Fig. 3.1. One distribution point is connected via fiber to the central office. A copper cable bundle runs from the DPU at the distribution point to the customer premises equipment of each subscriber. Crosstalk is created within the common section of the cable bundle. From the DP to the CPEs, the cable bundle branches out towards the individual subscribers such that some of them are close to the DPU while others are far away. This is referred to as non co-located topology.[1]

For downstream direction, signals of multiple users are jointly precoded while no receiver cooperation at the CPE side is possible, as they are physically at different places. MIMO downstream transmission optimization is covered in Sect. 3.2 for linear precoding and Sect. 3.3 for Tomlinson–Harashima precoding, [1, 2].

In upstream direction, CPEs send the uplink signals independently and the receivers at DPU side perform joint equalization. This is equivalent to a MIMO multiple-access channel. The receiver optimization is discussed in Sect. 3.4.

---

[1] In contrast to that, all CPEs are placed at the same distance to the DPU in the co-located topology, which is often used for lab testing or simplified simulations.

R. Strobel, *Channel Modeling and Physical Layer Optimization in Copper Line Networks*, Signals and Communication Technology,
https://doi.org/10.1007/978-3-319-91560-9_3

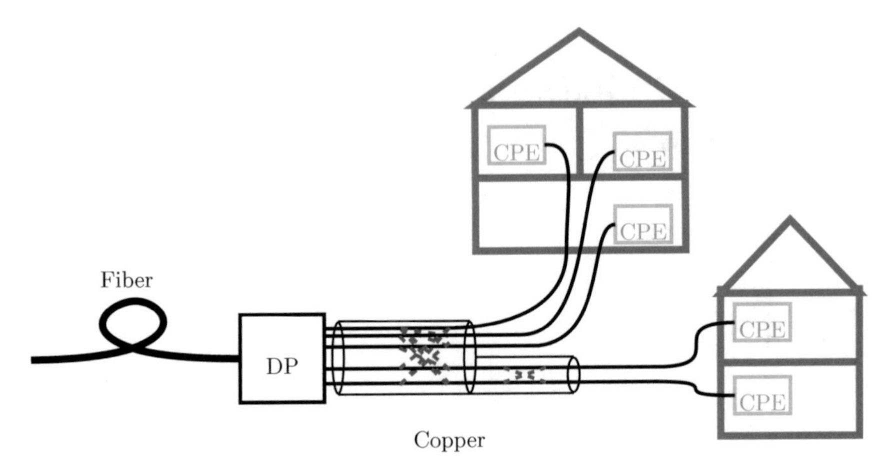

Fig. 3.1 Hybrid copper fiber network

Section 3.1 describes the signal processing components of a G.fast system, required for a single line point-to-point link. Section 3.6 discusses coexistence of G.fast with other legacy technologies [3].

Notation

The following notation is used throughout this chapter. Uppercase bold letters are used for matrices, lowercase bold letters are used for vectors and non-bold letters are used for scalars.

Whenever carrier-dependent parameters are used, they are marked with a discrete carrier index k according to $(.)^{(k)}$. For parameters depending on a discrete time or iteration index t, the notation $(.)^{[t]}$ is used. To address individual lines l of a multi-line system, the notation $(.)_l$ is used.

The Hermitian transpose of a matrix or a vector is written as $(.)^{\mathrm{H}}$ and the transpose is written as $(.)^{\mathrm{T}}$. When splitting a matrix $\boldsymbol{A}$ into individual column vectors, the notation $\boldsymbol{A} = [\boldsymbol{a}_1, \ldots, \boldsymbol{a}_L]$ is used, where $\boldsymbol{a}_l$ is the lth column of the matrix. Splitting a matrix $\boldsymbol{B}$ into individual row vectors is notated as $\boldsymbol{B} = [\boldsymbol{b}_1, \ldots, \boldsymbol{b}_L]^{\mathrm{T}}$, where $\boldsymbol{b}_l^{\mathrm{T}}$ is the lth row vector of the matrix $\boldsymbol{B}$.

$|\mathbb{I}|$ denotes the cardinality, the number of elements in the set $\mathbb{I}$, the operation diag (.) transforms a vector into a diagonal matrix with the corresponding diagonal elements and a matrix into a vector containing the diagonal elements of the matrix, tr () is the trace, the sum over the diagonal elements, of a matrix, $(.) \odot (.)$ is the element-wise product or Hadamard product of two matrices of the same dimensions, E [.] is the expectation operation. Gaussian distribution random vectors are defined as $\boldsymbol{n} \sim \mathcal{N}(\boldsymbol{\mu}, \boldsymbol{C}_{\mathrm{nn}})$ where $\boldsymbol{\mu}$ is the mean and $\boldsymbol{C}_{\mathrm{nn}}$ is the covariance matrix of $\boldsymbol{n}$, $\boldsymbol{I}_L$ is the identity matrix of size $L \times L$, $\mathbf{1}_L$ is the all-ones vector of size L and $\mathbf{0}_{L\times K}$ is the all-zeros matrix of size $L \times K$.

3.1 Transmission Model

This section gives a functional overview over G.fast point-to-point transmission, describing the modulation and coding method used for G.fast. Spectral constraints and transmit spectrum optimization for the single line case are discussed.

3.1.1 Transmitter-Receiver Model

The transmitter model for a single line G.fast PMD (physical media dependent) layer is shown in Fig. 3.2a. The general building blocks are similar for DMT-based (discrete multi tone) DSL technologies such as ADSL/ADSL2, VDSL 2 or G.fast. But in contrast to ADSL and VDSL2, G.fast uses time division duplexing (TDD) rather than frequency division duplexing (FDD) to separate upstream and downstream transmission.

The transmit signal is created in frequency domain with a number of K discrete subcarriers $k = 1, \ldots, K$. Most signal processing operations are done in the frequency domain while the discrete Fourier transform (DFT) is used for transformation between time domain signals and frequency domain signals. The PMD transmitter processing, as shown in Fig. 3.2a, starts with the trellis encoder and the QAM (quadrature amplitude modulation) modulator, which translates the data bits into QAM constellations with the appropriate number of bits $\hat{b}^{(k)}$ modulated onto each carrier k. It is discussed in more detail in Sect. 3.1.4.

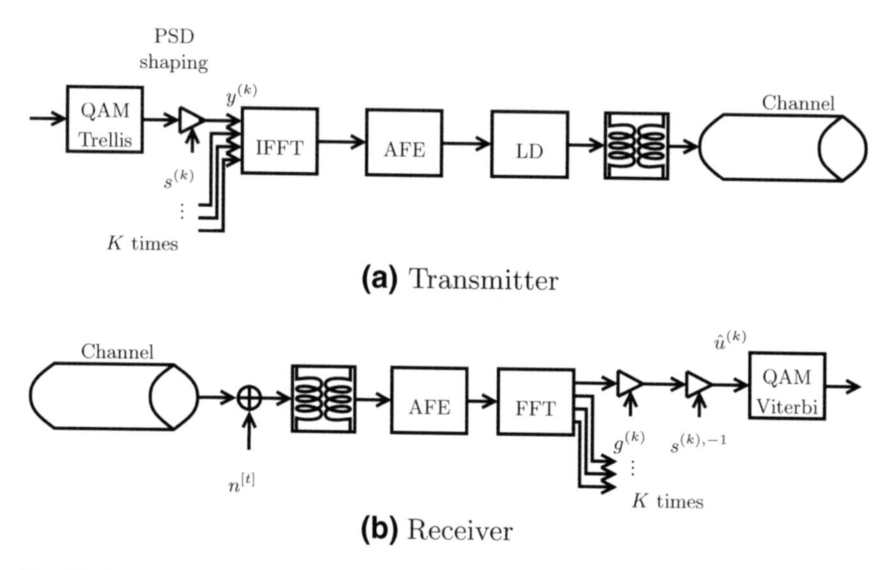

Fig. 3.2 Single line model

This is followed by the transmit spectrum shaping, where each carrier is independently scaled with the gain value $s^{(k)}$. The shaped signal is transformed into a time domain signal using inverse discrete Fourier transform (see Sect. 3.1.3). The cyclic extension is added and windowing is applied. The time domain digital samples are filtered and converted to analog signals in the analog front-end (AFE) as described in Sect. 3.1.2. This signal is amplified by the line driver (LD) and coupled onto the line using a transformer.

The receiver operates as shown in Fig. 3.2b. The noisy distorted receive signal is coupled into the receive AFE, where it is amplified and converted to digital samples. This is transformed to frequency domain, where the direct channel distortion is compensated by the adaptive frequency domain equalizer, which is a complex single-tap equalizer for each carrier $g^{(k)} \in \mathbb{C}$.

The transmitter gain shaping is inverted and the output signal is decoded by the QAM demodulator and Viterbi decoder. The output bits are processed by the higher layers.

Throughout the chapter, a frequency domain model as shown in Fig. 3.3 is used. It consists of K independent parallel channels. The channel is represented by one complex gain $H^{(k)}$ per carrier which represents the cable as well as the relevant transmit and receive filters, line driver and analog front-end components. The dependency between time domain channel and frequency domain channel for the DMT system is further discussed in Annex B.1.

The noise $n^{(k)} \sim \mathcal{N}(0, \sigma^{(k),2})$ is considered in frequency domain as complex Gaussian zero-mean noise with variance $\sigma^{(k),2}$. The noise variance per carrier is derived from the noise power spectral density (PSD) $\psi_{\mathrm{n}}(f)$ according to

$$\sigma^{(k),2} = \int_{f_k-\Delta f/2}^{f_k+\Delta f/2} \psi_{\mathrm{n}}(f)df \tag{3.1}$$

with a subcarrier spacing Δf.

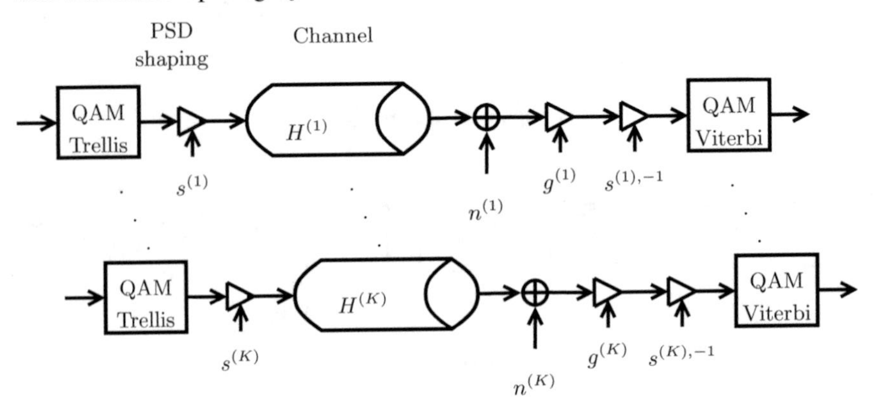

Fig. 3.3 Single line frequency domain model

The transmitter as well as the receiver is dividend into the components digital front-end (DFE) and analog front-end (AFE) as indicated in Fig. 3.2. The transmitter side also contains a power amplifier called line driver (LD).

3.1.2 *Analog Subsystem*

The transmit AFE consists of transmit filters and the digital-to-analog-converter (DAC). The filtered analog signal is amplified to the required transmit power by the line driver and then transmitted over the twisted pair line. A transformer is used to couple the line driver and the line.

The receive AFE contains analog channel equalizers, the analog gain control and the analog-to-digital converter (ADC). The limited ADC and DAC resolution are additional noise sources at the transmitter and receiver side which depend on corresponding signal amplitudes.

Quantization noise gives an upper bound on the number of bits per carrier. A similar effect is caused by the limited linearity of the line driver and other analog components.

The receive AFE is one of the noise sources of the G.fast system besides line noise and interference. Throughout this work a simplified analog subsystem model is assumed, which is based on the assumption that the analog components do not change the channel transfer function $H^{(k)}$ in frequency domain and the receiver noise floor is below the background noise PSD $\psi_{\mathrm{n}}(f)$. The linearity of transmitter and receiver is such that a maximum bit allocation b_{max} is achieved for any possible channel realization.

3.1.3 *DMT Modulation*

G.fast uses discrete multitone (DMT) modulation [4]. A DMT system performs signal processing mainly in frequency domain where the data is transmitted on K orthogonal subcarriers as indicated in Fig. 3.3. G.fast uses $K = 2048$ carriers for the 106 MHz profile and $K = 4096$ carriers for the 212 MHz profile. The subcarriers have a subcarrier spacing Δf, which is for G.fast $\Delta f = 51.75\,\mathrm{kHz}$.

The bit allocation and transmit power are configured per subcarrier in G.fast. This allows to adapt the transmitter settings to a frequency selective channel such as copper twisted pair wires and also to minimize impact of narrow-band interference or frequency selective noise. Bit allocation and transmit power per carrier are adjusted changes of the channel and noise conditions over time, using online reconfiguration methods discussed in Sect. 3.1.7.

The subcarriers are assumed to be orthogonal, even for the case of a distorted receive signal. This is achieved by the cyclic extension. It is selected such that the

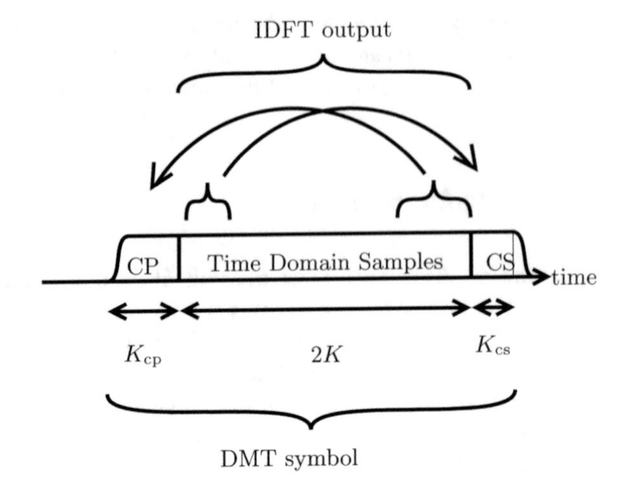

Fig. 3.4 DMT symbol with cyclic extension and windowing

length of the cyclic extension in time is greater or equal to length of the channel impulse response.

Figure 3.4 shows structure of a DMT symbol in time domain. It consists of the original inverse DFT (IDFT) samples which are extended by a cyclic prefix at the beginning of the symbol and a cyclic suffix at the end. Furthermore, the symbol boundaries are shaped with a raised cosine window [5] to reduce the out-of-band signal.

The cyclic extension adds some overhead. The number of DFT output samples is $2K$ which is extended by the number of Cyclic prefix samples K_{cp} and the number of cyclic suffix samples K_{cs} minus the windowing samples β. While the symbol time without cyclic extension would be $t_{\mathrm{sym}} = \frac{1}{\Delta f}$, the actual symbol time including the cyclic extension is

$$t_{\mathrm{sym}} = \frac{1}{2K\Delta f}(2K + K_{\mathrm{cp}} + K_{\mathrm{cs}} - \beta). \tag{3.2}$$

A standard configuration of 106 MHz G.fast is $K_{\mathrm{cp}} = 320$ and $K_{\mathrm{cs}} = \beta = 64$ or $K_{\mathrm{cs}} = \beta = 128$, which gives a cyclic extension overhead of 7.8125%. For the 212 MHz profile, $K_{\mathrm{cp}} = 640$ gives an equivalent symbol format with the same percentage of overhead. Both DMT symbol formats give a symbol time of $t_{\mathrm{sym}} = \frac{1}{48\,\mathrm{kHz}}$.

In opposite to OFDMA systems, where different users use only a fraction of the overall spectrum [6], each G.fast subscriber is assigned to the full transmit spectrum, even in a multi-line DPU. Some carriers on individual lines may be switched off due to a low channel quality.

3.1.4 QAM Modulation and Forward Error Correction

QAM modulation and forward error correction are important building blocks of the G.fast physical layer to guarantee efficient and reliable data transmission. The performance of nonlinear precoding depends on the forward error correction method due to the modulo operation which is required in the receiver [7].

Data is modulated on each subcarrier with a specific constellation size $\hat{b}^{(k)}$ for carrier k. This allows to adapt the data rate on each subcarrier to the available signal-to-noise ratio. There are constellations for 1 bit up to $b_{\max} = 12$ bits in standard G.fast [5].[2] VDSL2 allows constellations up to 15 bit [8]. The transmit spectrum is shaped with per-carrier gains to match the power constraints.

Figure 3.5 shows the constellations for 1–5 bits. Larger even constellations are created by replacing each constellation point of the 2 bit smaller constellation by four equally spaced constellation points, starting with the 4-QAM constellation. For odd constellations of 7 and more bits, the 5 bit constellation is used as a base constellation. The QAM modulator $\text{qam}_\text{e}(.)$ maps a bit vector $\boldsymbol{v}^{(k)} = \left[v_1^{(k)}, \ldots, v_{\hat{b}^{(k)}}^{(k)}\right], v_i^{(k)} \in \{0, 1\}$ to a complex constellation point $u^{(k)}$

$$u^{(k)} = \text{qam}_\text{e}\left(\boldsymbol{v}^{(k)}\right). \tag{3.3}$$

The QAM decoder $\text{qam}_\text{d}(.)$ maps a received signal $\hat{u}^{(k)}$ to the corresponding bit vector $\hat{\boldsymbol{v}}^{(k)}$ according to

$$\hat{\boldsymbol{v}}^{(k)} = \text{qam}_\text{d}\left(\hat{u}^{(k)}\right). \tag{3.4}$$

The mapping between a bit sequence $\hat{\boldsymbol{v}}^{(k)}$ and the corresponding constellation point for each constellation size $\hat{b}^{(k)}$ is defined in the G.fast standard [5]. For uncoded QAM modulation, the received signal is mapped to the closest point in terms of the squared distance

$$\text{qam}_\text{d}\left(\hat{u}^{(k)}\right) = \arg \min_{\boldsymbol{v}_1, \ldots, \boldsymbol{v}_{2^{\hat{b}}}} \left(\left|\hat{u}^{(k)} - \text{qam}_\text{e}\left(\boldsymbol{v}_i\right)\right|^2\right). \tag{3.5}$$

The 3-bit constellation has been redesigned in G.fast to reduce the power loss for nonlinear precoding [9]. The other constellations are the same in VDSL2 [10] and G.fast. In general, diamond-shaped constellations like the 3-bit G.fast constellation have advantages compared to the VDSL odd-bits constellations, especially for nonlinear precoding [11].

The achievable rate per carrier $b^{(k)}$ depends on the signal-to-noise ratio $SNR^{(k)}$ on the carrier. It is given by

$$b^{(k)} = \log_2\left(1 + \frac{SNR^{(k)}}{\Gamma}\right) \tag{3.6}$$

[2] An extension to 14 bit constellations is defined in an amendment to the G.fast standard.

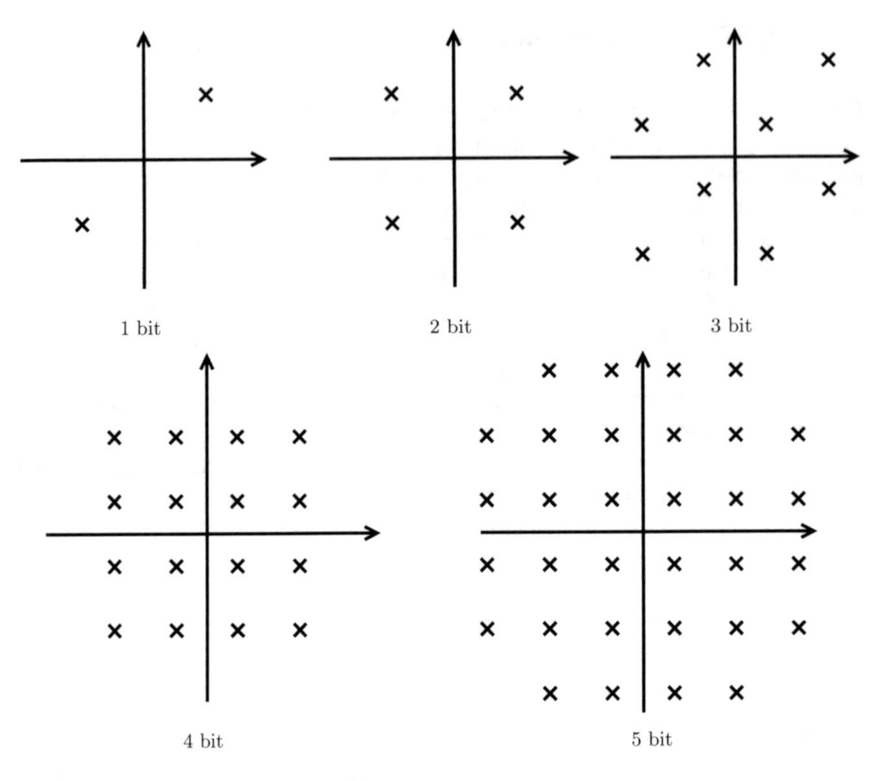

Fig. 3.5 G.fast constellations for 1–5 bit

with the SNR gap Γ [12]. For capacity of a Gaussian channel with Gaussian modulation, the SNR Gap is $\Gamma = 0\,\text{dB}$. For a system using coded QAM modulation, the SNR gap consists of three components $\Gamma = 9.8\,\text{dB} + \gamma_\text{m} - \gamma_\text{c}$ where 9.8 dB [12] is the SNR gap required to achieve the target bit error rate with uncoded modulation, γ_c is the coding gain for channel coding and γ_m is the SNR margin.

SNR margin γ_m is a parameter to increase robustness of the G.fast link against SNR changes. G.fast is designed for a target bit error rate (BER) of 10^{-7}, which is achieved $\gamma_\text{m} = 0\,\text{dB}$ SNR margin. Usually, $\gamma_\text{m} = 6\,\text{dB}$ is used to ensure a reliable operation of the link.[3]

The number of bits $\hat{b}^{(k)}$ to be modulated on a specific carrier depends on the available SNR. While Eq. (3.6) can be used as an upper bound, the bit allocation $\hat{b}^{(k)}$, allows only integer values between 1 and b_max. It is derived from the actual SNR per carrier $SNR^{(k)}$ using thresholds $SNR_{\text{req},\hat{b}}$ for each constellation. $SNR_{\text{req},\hat{b}}$ is the minimum SNR required to load $\hat{b}$ bits under the given bit error rate constraint. The required SNR values can be derived from Eq. (3.6), which gives

[3]For G.fast, 3 dB SNR margin are sometimes considered to be sufficient due to the reconfiguration methods such as seamless rate adaptation described in Sect. 3.1.7.

$$SNR_{\text{req},\hat{b}} = (2^{\hat{b}} - 1)\Gamma. \tag{3.7}$$

For the modulation and channel coding implementation of G.fast, the required SNR values differ from Eq. (3.7). Therefore, the corresponding SNR values $SNR_{\text{req},\hat{b}}$ are derived in simulations for each G.fast constellation, including the forward error correction scheme.

The bit allocation $\hat{b}^{(k)}$ on carrier k is selected with respect to the required SNR values $SNR_{\text{req},\hat{b}}$ to satisfy

$$\hat{b}^{(k)} = \arg\max_{\hat{b}} \text{ s.t. } \frac{SNR^{(k)}}{\gamma_{\text{m,min}}} \geq SNR_{\text{req},\hat{b}} \tag{3.8}$$

and achieve a certain minimum SNR margin $\gamma_{\text{m,min}} \geq 0\,\text{dB}$.

Simulation results and tables with the SNR values $SNR_{\text{req},\hat{b}}$ for G.fast are summarized in Appendix B.2 for uncoded QAM modulation, Appendix B.3 for trellis-coded modulation (TCM) and Appendix B.4 for trellis- and Reed-Solomon-coding.

Error Probabilities for Uncoded Linear and Modulo Receivers

The bit error rate for transmission of a certain QAM constellation over a channel with Gaussian noise depends on the noise variance σ^2 and the distance d between two constellation points. The bit error rate furthermore depends on whether a linear or a modulo receiver is used, which is explained in Fig. 3.6. For linear receivers, the constellation points on the boundaries have a lower decision error probability, as Fig. 3.6a indicates. For the modulo receiver, as in Fig. 3.6b, the error probability on boundary constellation points is the same as for the interior points.

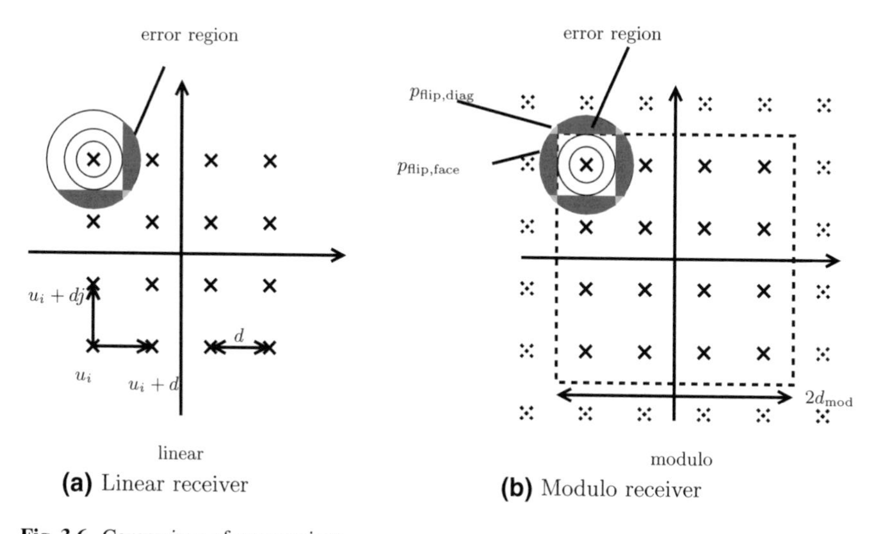

(a) Linear receiver **(b)** Modulo receiver

Fig. 3.6 Comparison of error regions

Modulo receivers perform a modulo operation mod (.) in advance to the QAM decoding to bring received constellation points into the area inside the modulo region, which is indicated in Fig. 3.6b for the 4 bit QAM constellation. Assuming a square modulo region of size $2d_{\text{mod}}$ by $2d_{\text{mod}}$ (see Fig. 3.6), the modulo operation on a constellation point u can be described as [13]

$$\mod(u) = u - 2d_{\text{mod}} \left(\left\lfloor \frac{\text{Re}\{u\}}{2d_{\text{mod}}} + \frac{1}{2} \right\rfloor + j \left\lfloor \frac{\text{Im}\{u\}}{2d_{\text{mod}}} + \frac{1}{2} \right\rfloor \right). \tag{3.9}$$

Figures B.6 and B.7 in Annex B.2 show that the SNR gap approximation with 9.8 dB is not exact, especially for the modulo receiver. Values for the required SNR $SNR_{\text{req},\hat{b}}$ of each constellation $\hat{b}$ are derived according to

$$SNR_{\text{req},\hat{b}} = \min SNR \text{ s.t. } p_{\text{error},\hat{b}}(SNR) \leq p_{\text{error,target}} \tag{3.10}$$

from the analytical expression for bit error probability $p_{\text{error},\hat{b}}(SNR)$ as given by Eq. B.13 in Appendices B.13 or from a bit error rate simulation for the target bit error rate $p_{\text{error,target}}$, e.g, $p_{\text{error,target}} = 10^{-7}$.

While for larger constellations, the BER of the linear receiver and the modulo receiver are almost identical at a given SNR, there is a gap for the small constellations, which is the modulo loss [14]. Tables B.2 and B.3 in Annex B.2 list the SNR values $SNR_{\text{req},\hat{b}}$ for G.fast constellations up to $\hat{b} = 14$ bit. For simulated data rates in Chaps. 3 and 4, Eqs. (3.8) and (3.10) are used while the analytical rate expression of Eq. (3.7) gives a good approximation of the achievable data rates and is used as a rate optimization objective.

Trellis-Coded Modulation

G.fast uses forward error correction for a more efficient data transmission. Simulation results show that trellis coded modulation behaves different for linear and modulo receivers [15]. The coding scheme used is the 4D Wei code [16].

At the transmitter, only two least significant bits of each carrier are processed by the trellis encoder [5], which takes 3 bit from the input data stream and adds one bit of redundancy to a pair of subcarriers.[4] The number of overhead bits per DMT symbol K_{oh} is given by

$$K_{\text{oh}} = \left\lceil \frac{K_{\text{used}} - K_{\text{single}}/2}{2} - 4 \right\rceil \tag{3.11}$$

where K_{used} is the number of used tones with bit allocation of 1 or more bits and K_{single} is the number of 1 bit carriers [5]. The trellis code gives a coding gain around $\gamma_{\text{c}} \approx 5.2$ dB. As indicated in Annex B.3, the coding gain is not the same for each constellation.

[4] One-bit carriers are a special case. A pair of 1-bit carriers is grouped to a 2-bit constellation in advance to trellis coding to keep the coding scheme for this special case. Therefore, the number of 1-bit carriers must be even.

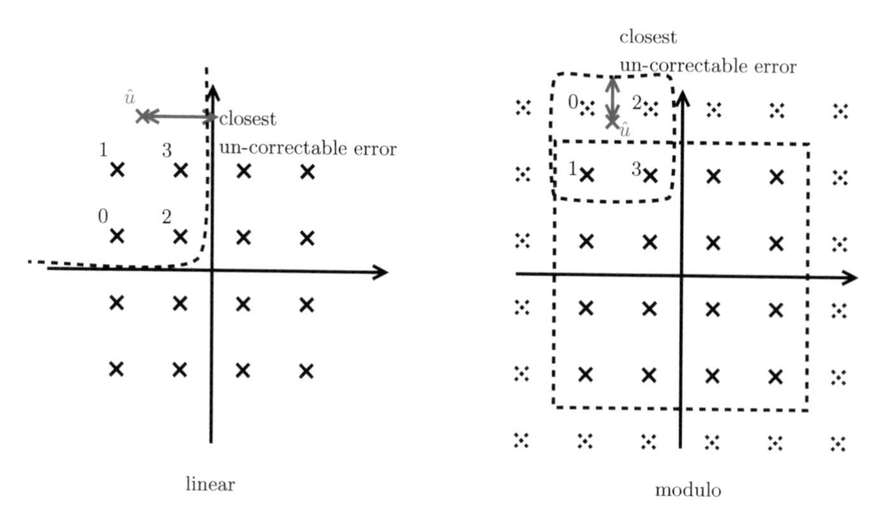

Fig. 3.7 Coded modulation with linear and modulo receivers

At the receiver side, the Viterbi decoder selects one out of four constellation point which are closest to the received constellation point $\hat{u}$ (see Fig. 3.7). The Viterbi decoder can only decode the received signal correctly if the received constellation point is within a certain region around the correct constellation point, which is indicated by the dashed line in Fig. 3.7. As shown in Fig. 3.7, the Viterbi decoder operation is different for linear and modulo receivers, because the procedure to select the constellation points which are closest to the received signal is different. Therefore, the modulo loss is also present for trellis coded modulation.

Comparing the modulo loss from Tables B.4 and B.5 in Appendix B.3 with the corresponding tables in Appendix B.2, the modulo loss in the trellis-coded system is even higher than for the uncoded system, which is mainly due to the lower SNR required to achieve the target bit error rate when using trellis coding [7].

Figures B.8 and B.9 in Annex B.3 show the SNR versus BER simulations for trellis coded modulation on the G.fast system. The required SNR values $SNR_{\mathrm{req},\hat{b}}$ for each G.fast constellation are summarized in Table B.4 for the linear receiver and Table B.5 for the modulo receiver.

Tone Ordering and Margin Equalization

The bit allocation in G.fast is selected according to Eq. (3.8) for a given target SNR margin $\gamma_{\mathrm{m,min}}$ between the actual SNR and the required SNR, $SNR_{\mathrm{req},\hat{b}^{(k)}}$. As only integer values for $\hat{b}$ can be selected, the actual margin $\gamma_{\mathrm{m}}^{(k)}$ as given by

$$\gamma_{\mathrm{m}}^{(k)} = SNR^{(k)} - SNR_{\mathrm{req},\hat{b}^{(k)}} \tag{3.12}$$

is different for each carrier and higher than the target margin. The excess SNR margin causes a loss of data rate of $\frac{1}{2}$ bit per carrier on average, compared to capacity according to Eq. (3.6). The properties of trellis coded modulation can be used to overcome the issue and benefit from the excess margin.

The 4D Wei code [16] used in G.fast operates on tone pairs, which are jointly encoded. The tones are not processed in the natural order from the first to the last tone, but according to a configurable tone ordering table (TOT, [5]). It is a vector of tone indices $\boldsymbol{o} = \left[o^{(1)}, \ldots, o^{(K)}\right]^{\mathrm{T}}$ which defines tone indices in the order of trellis processing.

The tone ordering can be selected such that a carrier with high margin and a carrier with low margin form a tone pair. This is achieved by constructing the tone ordering table $\boldsymbol{o}$ according to

$$o^{(k)} = \begin{cases} \arg \min\limits_{i:i \notin o_1, \ldots, o^{(k-1)}} \gamma_{\mathrm{m}}^{(i)} & \text{for } k \text{ even} \\ \arg \max\limits_{i:i \notin o_1, \ldots, o^{(k-1)}} \gamma_{\mathrm{m}}^{(i)} & \text{for } k \text{ odd.} \end{cases} \quad (3.13)$$

With that, the target bit error rate is achieved at a lower target margin $\gamma_{\mathrm{m,min}}$, which can be approximated by

$$\gamma_{\mathrm{m,min}} = \min_{i \in 1, \ldots, T, i \text{ odd}} \sqrt{\gamma_{\mathrm{m}}^{(o_i)} \gamma_{\mathrm{m}}^{(o_{i+1})}}. \quad (3.14)$$

The increase of SNR margin can be translated in a higher bitloading and higher data rates [17].

An alternative way of margin equalization is to use the transmit gain scaling $s^{(k)}$ to reduce the transmit power such that the SNR margin on all carriers is equal to the target margin. Depending on the spectral constraints, transmit power may also be increased on some carriers to achieve higher bit loading and a higher data rate.[5]

Reed–Solomon Coding and Interleaving

The G.fast link is additionally protected by a Reed–Solomon (RS) code. The code block size and overhead are configurable. The code block size in bytes is $N_{\mathrm{rs}} = 32, \ldots, 255$ which consists of K_{rs} data bytes and $R_{\mathrm{rs}} = 2, \ldots, 16$ redundancy bytes. This RS code is further analyzed in terms of the BER performance.

The Reed–Solomon code is used to correct bit errors in the unprotected part of the QAM constellations as well as error bursts caused by the trellis code due to wrong decisions.

The largest supported RS code with $K_{\mathrm{rs}} = 239$ data bytes and $R_{\mathrm{rs}} = 16$ check bytes, a 239/255 RS code, can correct up to 8 erroneous bytes. When large constellation sizes are used, especially with 8 and more bits per carrier, one trellis error burst may create more than 8 byte failures which exceeds the correction capabilities of the largest overhead configuration with $R_{\mathrm{rs}} = 16$.

[5]This is not possible for G.fast downstream direction. Due to the precoding, the transmit gain scaling is controlled by the transmitter side. In VDSL2 as well as in the G.fast upstream, margin equalization by power allocation is possible. The method is explained briefly in Sect. 3.1.7.

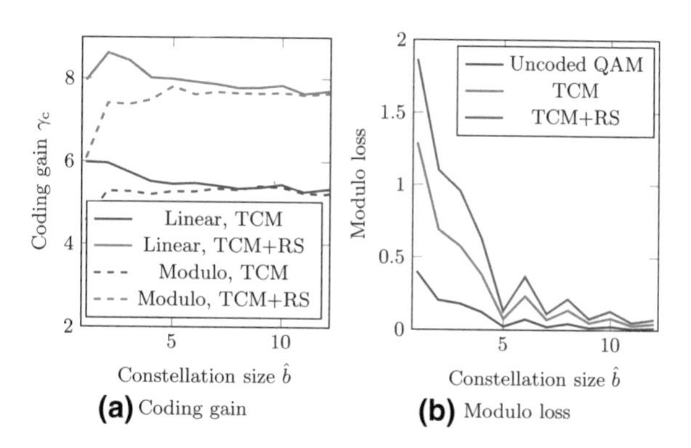

Fig. 3.8 Coding gain and modulo loss for the coding schemes applied in G.fast, trellis coded modulation (TCM) and trellis with Reed–Solomon coding, using a 239/255-Reed–Solomon code as an example

To overcome that, interleaving over multiple RS code words is performed. It is possible to group up to 16 RS code blocks to a data transmission unit (DTU) and do interleaving within the DTU. This is called intra-DTU interleaving and allows to correct most of the trellis burst errors. Figures B.10 and B.11 in Appendix B.4 show the bit error rate as a function of the SNR gap to capacity for Reed Solomon and trellis coding with intra-DTU interleaving. The optimal settings of the Reed–Solomon code configuration values K_{rs} and R_{rs} are selected for an optimal trade-off between overhead and coding gain.

Again, the required SNR for each constellation size is shown in Tables B.6 and B.7 for the linear as well as for the modulo case. The combined coding gain of trellis coding, Reed–Solomon coding and intra-DTU interleaving is around $\gamma_c \approx 7.6\,\text{dB}$.

To conclude the analysis of G.fast modulation and coding and its dependency to the precoding method, the coding gain γ_c as well as the modulo loss are shown for trellis coded modulation and for the combination of trellis and Reed–Solomon coding in Fig. 3.8. The coding gain, as shown in Fig. 3.8a for trellis coded modulation as well as for trellis and Reed–Solomon coding, depends on the constellation size. Especially for the small constellations with 1 and 2 bit, the coding gain achieved by the modulo receiver is lower than the coding gain for a linear receiver. The modulo loss, the difference between the coding gain of linear and modulo receiver, is shown in Fig. 3.8b. It is observed that the modulo loss increases with increasing coding gain, i.e., the Reed–Solomon and trellis coding scheme shows a higher modulo loss than trellis coding only or uncoded QAM modulation. Furthermore, the modulo loss decreases with increasing constellation size $\hat{b}$. In general, the constellations with different constellation shapes behave different, i.e., the odd-bit constellations show a lower modulo loss than the even-bit constellations.

3.1.5 Data Rate Calculation

Performance comparisons in this work are based on the achieved data rates. To derive the actual data rate of a link from a certain SNR or bit allocation per carrier, losses due to framing and coding are considered. In a communication system such as G.fast, different data rates are observed at the interfaces between different layers. Referring to the G.fast reference model as shown in Figs. 1.3 and 1.4, the data rate definition used in this work is the data rate at the γ interface between layer 1 and layer 2, which takes into account all overhead contributions of the G.fast physical layer. This is compared to the line capacity.

The line capacity for a single G.fast line is given by

$$R = \Delta f \left(\sum_{k=1}^{K} b^{(k)} \right) \tag{3.15}$$

with $b^{(k)}$ according to Eq. (3.6), using a SNR gap $\Gamma = 0\,\text{dB}$ and a subcarrier spacing Δf. The actual data rate of a link is lower due to the overhead.

The number of modulated bits $\hat{b}^{(k)}$ per carrier k is determined by

$$\hat{b}^{(k)} = \max_{\hat{b}} \text{ s.t. } SNR_{\text{req},\hat{b}} \leq \frac{SNR^{(k)}}{\gamma_{\text{m,min}}}, \tag{3.16}$$

using the required SNR values $SNR_{\text{req},\hat{b}}$ from Appendices B.3, B.4 or Appendix B.2 according to the configured coding scheme. The SNR margin $\gamma_{\text{m,min}} \geq 1$ is used to increase robustness against changes of the channel or noise conditions. For DSL and G.fast applications, $\gamma_{\text{m,min}} = 6\,\text{dB}$ is a common setting [12].

The data rate of a link is given by the sum of modulated bits $\hat{b}^{(k)}$ where overhead is subtracted. Assuming that a DMT symbol is transmitted within a time t_{sym}, which includes the cyclic extension and windowing, the data rate is given by

$$R = \frac{\eta_{\text{code}}\eta_{\text{framing}}}{t_{\text{sym}}} \left(\sum_{k=1}^{K} \hat{b}^{(k)} - K_{\text{oh}} \right) - R_{\text{oh}}. \tag{3.17}$$

where $\eta_{\text{code}} = \frac{K_{\text{rs}}}{N_{\text{rs}}}$ is the code rate of the Reed–Solomon code, e.g., $\eta_{\text{code}} = 239/255$ and K_{oh} is the number of overhead bits per DMT symbol added by the trellis code according to Eq. (3.11). Additional overhead η_{framing} is caused by the TDD framing which is discussed in Sect. 4.1. For a usual frame configuration, $\eta_{\text{framing}} = \frac{7(M_{\text{F}}-1)+(M_{\text{F}}-3)}{8M_{\text{F}}} = 96.5\%$ with $M_{\text{F}} = 36$ holds. R_{oh} accounts for the data rate required for overhead channels.

The ratio between upstream and downstream is configurable. Therefore, the aggregated (upstream + downstream) data rate is used for performance comparison. To compare different downstream precoding schemes, 100% downstream transmission is assumed.

3.1.6 Power Constraints and Spectrum Shaping

Two power constraints must be satisfied for G.fast transmission [18]. One is the power spectral density (PSD) of the signal on the line which is limited by a PSD mask $\psi_{\text{mask}}(f)$, defined by regulation. This translates into a power limit $p_{\text{mask}}^{(k)} \in \mathbb{R}$ per subcarrier k according to

$$x^{(k)} = |s^{(k)}|^2 \leq p_{\text{mask}}^{(k)} \tag{3.18}$$

which shall be satisfied on the line. The spectral mask for 106 MHz G.fast is shown in Fig. 3.9 together with an example of an actual transmit PSD. Hereby, $x^{(k)} = |s^{(k)}|^2$ is the actual transmit power on carrier k. The maximum power per carrier k is derived from the spectral mask $\psi_{\text{mask}}(f)$ according to

$$p_{\text{mask}}^{(k)} = \int_{f_k-\Delta f/2}^{f_k+\Delta f/2} \psi_{\text{mask}}(f)df\,. \tag{3.19}$$

The spectral mask is not only defined in the transmit band, where it is controlled with the transmit gains $s^{(k)}$, but also out-of-band. The out-of-band transmit PSD is formed with transmit filters.

The second constraint comes from limited capabilities of the transmit amplifier. It gives a maximum sum-power limit p_{sum} to be

$$\sum_{k=1}^{K} x^{(k)} \leq p_{\text{sum}} \tag{3.20}$$

for a system with K subcarriers.

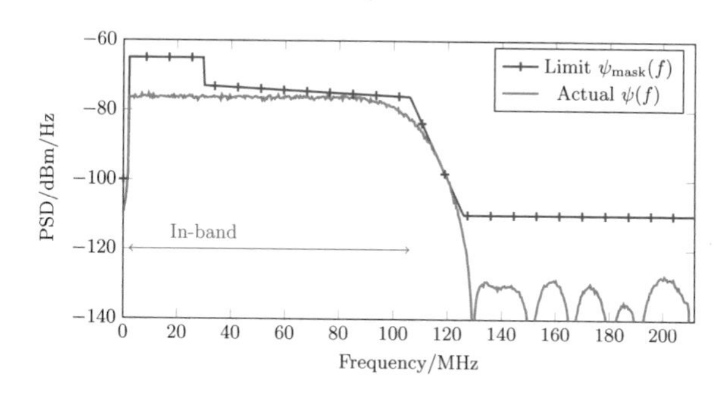

Fig. 3.9 In-band and out-of-band spectrum for 106 MHz G.fast

Optimization of the in-band transmit spectrum is based on a water-filling approach, which is extended to support the spectral mask constraint, Eq. (3.18), in addition to the sum-power constraint. Optimization is done for the per-carrier power values $x^{(k)}$ rather than for the gain values $s^{(k)}$, because only the power values occur in the spectral constraints Eqs. (3.18) and (3.20) as well as in the objective function. The objective is the data rate, approximated by the continuous function $\sum_{k=1}^{K} \min\left(\log_2\left(1+\frac{|H^{(k)}|^2 x^{(k)}}{\sigma^{(k),2}\Gamma}\right), b_{\max}\right)$ and taking the maximum bit allocation $b_{\max}$ into account.

Assuming a background noise level $\sigma^{(k),2}$, a direct channel transfer function of $H^{(k)}$ on carrier k and the SNR gap Γ, the following optimization problem is given for rate maximization on the single G.fast line with k carriers

$$\min_{x^{(k)},k=1,\dots,K} \sum_{k=1}^{K} -\min\left(\log_2\left(1+\frac{|H^{(k)}|^2 x^{(k)}}{\sigma^{(k),2}\Gamma}\right), b_{\max}\right) \tag{3.21}$$

$$\text{s.t.} \sum_{k=1}^{K} x^{(k)} \le p_{\text{sum}} \quad x^{(k)} \le p_{\text{mask}}^{(k)} \quad x^{(k)} \ge 0 \tag{3.22}$$

Due to the bit loading upper bound, the objective is continuous, but not differentiable. To avoid that, the bit loading upper bound is reformulated as another power constraint according to

$$\min_{x^{(k)},k=1,\dots,K} \sum_{k=1}^{K} -\log_2\left(1+\frac{|H^{(k)}|^2 x^{(k)}}{\sigma^{(k),2}\Gamma}\right) \tag{3.23}$$

$$\text{s.t.} \sum_{k=1}^{K} x^{(k)} \le p_{\text{sum}} \tag{3.24}$$

$$\text{s.t. } x^{(k)} \le \min(p_{\text{mask}}^{(k)}, p_{\text{bmax}}^{(k)}) = \bar{p}_{\text{mask}}^{(k)} \tag{3.25}$$

$$\text{s.t. } x^{(k)} \ge 0 \tag{3.26}$$

where the bit loading upper bound in incorporated into the spectral mask constraint, Eq. (3.25), and $p_{\text{bmax}}^{(k)}$ is the power required for the maximum bit loading on carrier k. It is derived with respect to the receiver noise level to be

$$p_{\text{bmax}}^{(k)} = (2^{b_{\max}} - 1)\frac{\sigma^{(k),2}\Gamma}{|H^{(k)}|^2}. \tag{3.27}$$

Replacing the bit loading constraint $b^{(k)} \leq b_{\max}$ with the power constraint of (3.25) and Eq. (3.27) gives a convex optimization problem. The resulting data rate is the same, while the use of (3.27) guarantees that the resulting optimal spectrum has the minimum transmit power required. The maximum bit loading constraint and the spectral mask constraint are merged in Eq. (3.25).

The Lagrange function $\phi(x^{(k)}, \mu_{\text{sum}}, \mu_{\text{mask}}^{(k)}, \mu_0^{(k)})$ is given by

$$\phi(x^{(k)}, \mu_{\text{sum}}, \mu_{\text{mask}}^{(k)}, \mu_0^{(k)}) = \sum_{k=1}^{K} -\log_2\left(1 + \frac{\left|H^{(k)}\right|^2 x^{(k)}}{\sigma^{(k),2}\Gamma}\right)$$
$$+ \mu_{\text{sum}}\left(\sum_{k=1}^{K} x^{(k)} - p_{\text{sum}}\right) + \sum_{k=1}^{K} \mu_{\text{mask}}^{(k)}\left(x^{(k)} - \bar{p}_{\text{mask}}^{(k)}\right) - \sum_{k=1}^{K} \mu_0^{(k)} x^{(k)} \tag{3.28}$$

with the Lagrange multipliers μ_{sum} for the sum-power constraint, $\mu_0^{(k)}$ for positiveness and $\mu_{\text{mask}}^{(k)}$ for the combined spectral mask and bit loading constraint.

The dual feasibility condition for the optimization problem is

$$\frac{\partial\phi(x^{(k)}, \mu_{\text{sum}}, \mu_{\text{mask}}^{(k)})}{\partial x^{(k)}} = -\frac{1/(\ln 2)}{\left|H^{(k)}\right|^{-2}\sigma^{(k),2}\Gamma + x^{(k)}}$$
$$+ \mu_{\text{sum}} + \mu_{\text{mask}}^{(k)} - \mu_0^{(k)} = 0 \forall\, k = 1, \ldots, K. \tag{3.29}$$

The Lagrange multiplier of the sum-power constraint μ_{sum} is derived from Eq. (3.29) by a water-filling approach for a given set of carriers $\mathbb{I}_{\text{mask}} \subseteq \{1, \ldots, K\}$ where spectral mask constraint (Eq. (3.25)) is active and a given set $\mathbb{I}_0 \subset \{1, \ldots, K\}$, where the positiveness constraint is active (Eq. (3.26)). For the remaining carriers $\mathbb{I}_{\text{fill}} = \{i : i \notin \mathbb{I}_{\text{mask}} \wedge i \notin \mathbb{I}_0\}$, only the sum-power constraint is active.

For the Lagrange multiplier μ_{sum},

$$\frac{1}{\mu_{\text{sum}}} = \frac{1}{|\mathbb{I}_{\text{fill}}|}\left(p_{\text{sum}} + \sum_{k\in\mathbb{I}_{\text{fill}}} \frac{\Gamma\sigma^{(k),2}}{\left|H^{(k)}\right|^2} - \sum_{k\in\mathbb{I}_{\text{mask}}} \bar{p}_{\text{mask}}^{(k)}\right) \tag{3.30}$$

holds, which follows from Eq. (3.29) because for carriers $k \in \mathbb{I}_{\text{sum}}$, $\mu_0^{(k)} = 0$ and $\mu_{\text{mask}}^{(k)} = 0$, $\frac{1}{\mu_{\text{sum}}} = x^{(k)} + \frac{\Gamma\sigma^{(k),2}}{|H^{(k)}|^2}$ must be satisfied.

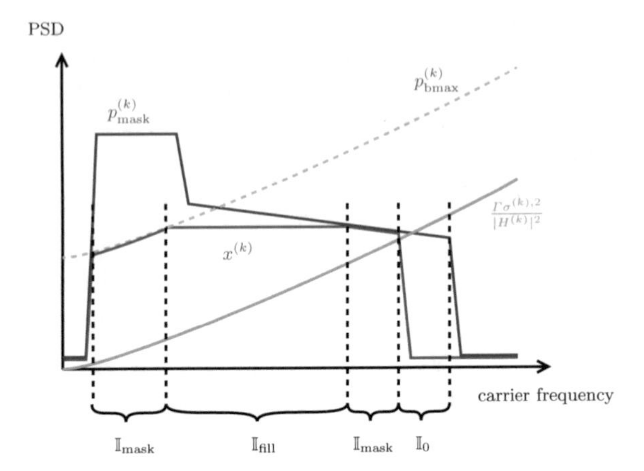

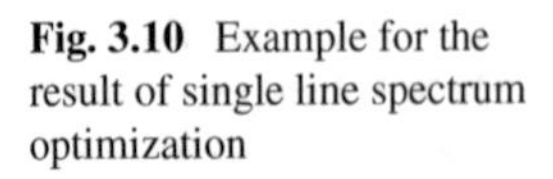
Fig. 3.10 Example for the result of single line spectrum optimization

The optimal power allocation with respect to μ_{sum} is given by

$$x^{(k)} = \begin{cases} 0 & \text{for } k \in \mathbb{I}_0 \\ \bar{p}_{\mathrm{mask}}^{(k)} & \text{for } k \in \mathbb{I}_{\mathrm{mask}} \\ \frac{1}{\mu_{\mathrm{sum}}} - \frac{\Gamma\sigma^{(k),2}}{|H^{(k)}|^2} & \text{otherwise,} \end{cases} \tag{3.31}$$

which satisfies the primal feasibility, dual feasibility and complementary slackness conditions when the sets $\mathbb{I}_0$ and $\mathbb{I}_{\mathrm{mask}}$ are selected correctly. The search for $\mathbb{I}_0$ and $\mathbb{I}_{\mathrm{mask}}$ is performed according to Algorithm 1.[6] Figure 3.10 illustrates the result for an example that may occur on a G.fast line. At lower frequencies, the bit loading upper bound is active due to the low channel attenuation. At higher frequencies, the sum-power constraint is active until the spectral mask is limiting the transmit power $x^{(k)}$ again due to the reducing spectral mask at higher frequencies. Some of the highest carriers may have $x^{(k)} = 0$ due to the increasing channel attenuation at high frequencies.

Algorithm 1 summarizes the spectrum optimization algorithm. The rate optimization for a single G.fast link, as shown in this section, is the basis for optimization of multi-line systems.

[6] A full search over all possible combinations of the sets is not required, because the carriers with the highest noise level $\frac{\Gamma\sigma^{(k),2}}{|H^{(k)}|^2}$ are the first to be part of the 0-subcarriers $\mathbb{I}_0$ while the carriers with the smallest distance between noise level $\frac{\Gamma\sigma^{(k),2}}{|H^{(k)}|^2}$ and the effective spectral mask $\bar{p}_{\mathrm{mask}}^{(k)}$ are the first to be part of the mask limited carriers.

Algorithm 1 Transmit spectrum optimization for a single G.fast line

```
Calculate p̄_mask^(k)
Initialize I_0,init = {i : Γσ^(k),2/|H^(k)|^2 ≥ p̄_mask^(k)}, K_0 = |I_0|
Initialize I_mask = {}, K_mask = 0
Initialize I_fill
if K_0,init = K then
  x^(k) = 0 ∀k = 1, ..., K
else
  if Σ_{k∈I_fill} p̄_mask^(k) ≤ p_sum then
    x^(k) = p̄_mask^(k) ∀ k ∉ I_0,init; 0 otherwise
  else
    Sort noise values Γσ^(k),2/|H^(k)|^2 for carriers k ∉ I_0,init in descending order into vector
    k_noise
    for K_0 = 0 to K − 1 − K_0,init do
      Set I_0 = I_0,init ∪ {i : i ∈ k_noise,1, ..., k_noise,K_0}
      Sort difference values p̄_mask^(k) − Γσ^(k),2/|H^(k)|^2 for all k ∉ I_0 in ascending order into
      vector k_mask
      for K_mask = 0 to K − K_0 − K_0,init do
        Set I_mask = {i : i ∈ k_mask,1, ..., k_mask,K_mask}
        Set I_fill accordingly
        Evaluate Eq. (3.30) for μ_sum
        Evaluate Eq. (3.31) for power allocation
        if μ_sum > 0 and x^(k) > 0∀k ∈ I_fill and x^(k) ≤ p̄_mask^(k) ∀k ∈ I_fill then
          This is the solution, stop.
        end if
      end for
    end for
  end if
end if
```

3.1.7 *Online Reconfiguration*

Noise and channel conditions may change on an active G.fast link after initialization. Online reconfiguration mechanisms are an integral part of the G.fast physical layer to guarantee a stable operation of the link at the optimal data rate. The different rate adaptation mechanisms are

- **bit swap** to maintain the minimum SNR margin $\gamma_{\text{m,min}}$ on all carriers without changing the data rate,
- **seamless rate adaptation (SRA)** to adjust the data rate to the current SNR,
- **transmitter initiated gain adjustment (TIGA)** to change the transmit spectrum for downstream from the transmitter side,
- **fast rate adaptation (FRA)** to keep the link alive in case of a severe drop of SNR and
- **RMC parameter adjust (RPA)** to change the settings of the RMC overhead channel.

This subsection focuses on SRA and bit swap because TIGA is only required for multi-line G.fast systems [5] and FRA and RPA are not related to performance optimization. SRA as well as bit swap are triggered when the SNR margin per carrier $\gamma_{\mathrm{m}}^{(k)}$ exceeds certain thresholds. Both methods are part of the VDSL2 standard [10], also, while some changes were required to use the reconfiguration methods in G.fast [19, 20].

A bit swap is initiated when the SNR margin on any carrier is below a threshold $\gamma_{\mathrm{m}}^{(k)} < \gamma_{\mathrm{BS,min}}$ or above threshold $\gamma_{\mathrm{m}}^{(k)} > \gamma_{\mathrm{BS,max}}$ to guarantee $\gamma_{\mathrm{BS,min}} \leq \gamma_{\mathrm{m}}^{(k)} \leq \gamma_{\mathrm{BS,max}}\ \forall k = 1, \ldots, K$. Bit swap can be performed with low effort, because the data rate remains constant and as a result, the reconfiguration is not visible on the higher transmission layers, but only on the PMD sublayer (see Fig. 1.3).

SRA may change the data rate on the line, but it is only triggered when the average margin exceeds the SRA thresholds and violates the condition $\gamma_{\mathrm{SRA,min}} \leq \frac{1}{K}\sum_{k=1}^{K} \gamma_{\mathrm{m}}^{(k)} \leq \gamma_{\mathrm{SRA,max}}$. In both cases, a new bit allocation according to Eq. (3.16) is calculated and results in a new bit allocation $\hat{b}^{(k),[t+1]}$. For the bit swap, the number of bits must be kept constant before and after the bit swap, $\sum_{k=1}^{K} \hat{b}^{(k),[t]} = \sum_{k=1}^{K} \hat{b}^{(k),[t+1]}$ which is achieved by successively performing

$$\hat{b}^{(k_{\gamma,\mathrm{min}}),[t+1]} = \hat{b}^{(k_{\gamma,\mathrm{min}}),[t]} - 1 \tag{3.32}$$

with $k_{\gamma,\mathrm{min}} = \arg\min_{k=1,\ldots,K,\hat{b}^{(k),[t]}>0} \gamma_{\mathrm{m}}^{(k)}$ in case that the number of bits is too high. In case that the number of bits is too low,

$$\hat{b}^{(k_{\gamma,\mathrm{max}}),[t+1]} = \hat{b}^{(k_{\gamma,\mathrm{max}}),[t]} + 1 \tag{3.33}$$

with $k_{\gamma,\mathrm{max}} = \arg\max_{k=1,\ldots,K,\hat{b}^{(k),[t]}<b_{\mathrm{max}}} \gamma_{\mathrm{m}}^{(k)}$ is performed.

SRA as well as bit swap do not achieve the target SNR margin $\gamma_{\mathrm{m,min}}$ exactly because a change of the bit allocation by one bit results in approximately 3 dB change of the SNR margin. One method for margin equalization, which is based on a trellis coding, is discussed in Sect. 3.1.4. An alternative solution is to change the per-carrier transmit power $x^{(k)}$ with respect to the SNR margin. While VDSL2 allows to increase and decrease the transmit PSD with SRA or bit swap, G.fast only allows to reduce the transmit PSD. Therefore, the actual margin must be greater or equal to the target margin $\gamma_{\mathrm{m}}^{(k)} \geq \gamma_{\mathrm{m,min}}$. The transmit power $x^{(k)}$ is then updated from time step t to $t+1$ according to

$$x^{(k),[t+1]} = x^{(k),[t]} \frac{\gamma_{\mathrm{m,min}}}{\gamma_{\mathrm{m}}^{(k)}}. \tag{3.34}$$

The described online reconfiguration methods, bit swap and SRA, are initiated by the receiver side. In case of downstream precoding, transmitter initiated changes of the bit allocation and the transmit power are required to optimize performance.

3.2 Downstream Linear Precoding

MIMO signal processing in the G.fast downstream direction is used to mitigate performance losses due to crosstalk. Linear precoding methods combine several advantages to address the problem. They can be implemented with low complexity and there is practical experience from VDSL2 crosstalk cancelation [8]. VDSL2-based crosstalk cancelation mostly relies on zero-forcing precoding, as this is near-optimal for the channel conditions at frequencies below 30 MHz [21]. The transmit spectrum and the precoding are handled independently, as the impact of precoding on the transmit spectrum is negligible.

This is no longer the case for G.fast, as the channel conditions are different at high frequencies (see Chap. 2). More advanced strategies for crosstalk cancelation and spectrum optimization are required.

3.2.1 Downstream System Model

This section focuses on precoding and spectrum optimization for the G.fast downstream. The system model from Fig. 3.3 is extended to a MIMO description for precoding over L lines and k carriers.

The corresponding block diagram for one subcarrier k is shown in Fig. 3.11. MIMO transmit signal processing starts with the gain-scaling diagonal matrix $\boldsymbol{S}^{(k)} = \mathrm{diag}(s_1^{(k)}, \ldots, s_L^{(k)}) \in \mathbb{R}^{L \times L}$ which is inverted at the receiver side. Between the gain scaling stages $s_l^{(k)}$ and the channel, the linear precoder $\boldsymbol{P}^{(k)} \in \mathbb{C}^{L \times L}$ at the

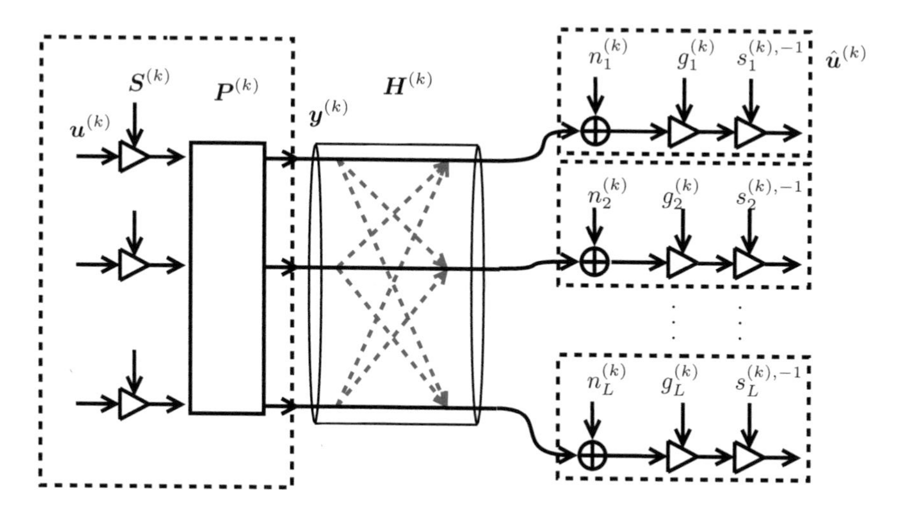

Fig. 3.11 Downstream system model representing one subcarrier

DPU is used to precompensate crosstalk between the lines. The MIMO channel for subcarrier k is described by the channel matrix $\boldsymbol{H}^{(k)} \in \mathbb{C}^{L \times L}$.

The receive equalizers of all lines are represented in the diagonal equalizer matrix $\boldsymbol{G}^{(k)} = \mathrm{diag}(g_1^{(k)}, \ldots, g_L^{(k)}) \in \mathbb{C}^{L \times L}$. The G.fast downstream transmission allows cooperation between the transmitters at the DPU, but not at the CPE side, because the CPEs are located at physically different places.

The system model is given by

$$\hat{\boldsymbol{u}}^{(k)} = \boldsymbol{S}^{(k),-1} \boldsymbol{G}^{(k)} \left(\boldsymbol{H}^{(k)} \boldsymbol{P}^{(k)} \boldsymbol{S}^{(k)} \boldsymbol{u}^{(k)} + \boldsymbol{n}^{(k)} \right) \tag{3.35}$$

for each subcarrier k. The transmit and receive signal vectors are $\boldsymbol{u}^{(k)} \in \mathbb{C}^L$ and $\hat{\boldsymbol{u}}^{(k)} \in \mathbb{C}^L$, respectively. The transmit signals are assumed to be statistically independent, zero-mean, unit power QAM signals. The receivers experience additive white Gaussian noise (AWGN) $\boldsymbol{n}^{(k)} \sim \mathcal{N}_{\mathbb{C}}(\boldsymbol{0}, \sigma^2 \boldsymbol{I})$.

The channel matrix $\boldsymbol{H}^{(k)}$ is square because as there is one transmitter at the DPU side connected to each receiver at the CPE side by one twisted pair of the cable binder.

3.2.2 MIMO Precoding Constraints

The G.fast power constraints are known from Sect. 3.1.6 for a single G.fast line. For MIMO transmission, the sum-power constraint translates into a per-line sum-power constraint and the spectral mask constraint into a per-line spectral mask constraint. There are no general sum power constraints for G.fast, as they are often assumed for wireless systems [22].

The transmit power constraint shall be satisfied for the precoder output signal $\boldsymbol{y}^{(k)} \in \mathbb{C}^L$ which is given by

$$\boldsymbol{y}^{(k)} = \boldsymbol{P}^{(k)} \boldsymbol{S}^{(k)} \boldsymbol{u}^{(k)}. \tag{3.36}$$

The per-line spectral mask constraint is given by

$$\mathrm{diag}\left(\boldsymbol{P}^{(k)} \boldsymbol{S}^{(k)} \boldsymbol{S}^{(k),\mathrm{H}} \boldsymbol{P}^{(k),\mathrm{H}} \right) \leq \boldsymbol{p}_{\mathrm{mask}}^{(k)} \quad \forall\, k = 1, \ldots, K \tag{3.37}$$

where the power constraints of individual lines are collected in a per-carrier power limit vector $\boldsymbol{p}_{\mathrm{mask}}^{(k)} \in \mathbb{R}^L$ for L lines.

The per-line sum-power constraint is given by

$$\sum_{k=1}^{K} \mathrm{diag}\left(\boldsymbol{P}^{(k)} \boldsymbol{S}^{(k)} \boldsymbol{S}^{(k),\mathrm{H}} \boldsymbol{P}^{(k),\mathrm{H}} \right) \leq \boldsymbol{p}_{\mathrm{sum}} \tag{3.38}$$

with the sum power vector $\boldsymbol{p}_{\mathrm{sum}} \in \mathbb{R}^L$.

3.2.3 Channel Estimation

There are two ways to perform channel estimation in G.fast. Pilot-based channel estimation with dedicated symbols with known content and blind channel estimation where the transmit signal is reconstructed at the receiver. In both cases, the receiver error $e_l^{(k),[t]}$ for line l, carrier k and time instance t is evaluated. For the pilot-based channel estimation, it is given by

$$e_l^{(k)[t]} = \hat{u}_l^{(k)[t]} - u_l^{(k),[t]} \tag{3.39}$$

with a known transmit signal $u_l^{(k),[t]}$. For blind channel estimation, the reference signal is reconstructed from the bit sequence $\hat{\boldsymbol{v}}_l^{(k),[t]}$ at the channel decoder output, which gives the receiver error signal

$$e_l^{(k)[t]} = \hat{u}_l^{(k)[t]} - \text{qam}_\text{e}\left(\hat{\boldsymbol{v}}_l^{(k),[t]}\right). \tag{3.40}$$

Assuming the target bit error rate of 10^{-7} or lower, the probability of estimation errors is low and the channel estimation is sufficiently accurate. The blind channel estimation, based on Eq. (3.40), allows frequent channel estimation updates at every transmitted symbol and is used, for example, to update the receive equalizer $g_l^{(k)}$ at the CPE side.

MIMO channel estimation is based on known transmit symbols as in Eq. (3.39), called sync symbols [5] . They are modulated with known 4-QAM constellations $u_{\text{sync}}^{(k)}$. For G.fast, depending on the frame format, usually every 288th symbol is a sync symbol.

Orthogonal sequences $w_l^{[t]}$ of length T are assigned to each line l, as described in Appendix B.6. The sync symbol at time instance t is given by

$$u_l^{(k)[t]} = u_{\text{sync}}^{(k)} w_l^{[t]}. \tag{3.41}$$

Over the length T sync symbols, sequences of different lines are mutually orthogonal and $\sum_{t=1}^{T} w_l^{[t]} w_m^{[t]} = 0$ holds for any $l \neq m$ [23]. One type of sequences to be used are Walsh–Hadamard sequences. Appendix B.6 gives more details on the construction of orthogonal matrices.

While for G.vector, the orthogonal sequence is built from $w_l^{[t]} \in \{-1, 1\}$, G.fast also allows zero-power symbols $w_l^{[t]} \in \{-1, 0, 1\}$ [23, 24]. For downstream precoder initialization and tracking, the receiver error $\boldsymbol{e}^{(k),[t]} = [e_1^{(k),[t]}, \ldots, e_L^{(k),[t]}]^\text{T}$ is reported from the CPEs to the DPU for each sync symbol using the vectoring feedback channel.[7] To reduce the data rate on the overhead channel, the error report may be compressed by interpolation over frequency [25].

[7] Alternatively, the received signal vector $\hat{\boldsymbol{u}}^{(k),[t]}$ can be reported over the vectoring feedback channel.

At the DPU side, the scaled channel estimation $\boldsymbol{H}_{\text{est}}^{(k)} \approx \boldsymbol{H}^{(k)}\boldsymbol{G}^{(k)}$ is derived from the reported error values $\boldsymbol{e}^{(k),[t]}$ according to[8]

$$\boldsymbol{H}_{\text{est}}^{(k)} = \frac{1}{T}\sum_{t=1}^{T} \boldsymbol{e}^{(k)[t]}\boldsymbol{u}^{(k)[t],\text{H}} + \boldsymbol{I}_L. \tag{3.42}$$

Channel estimation is performed initially when a new CPE is connected and continuously updated as long as the service is active.

3.2.4 Line Training and Joining

When a new CPE is connected to the network, a training sequence is executed to initialize certain transmission parameters and train the precoder and equalizer in downstream and upstream direction. As there may already be active lines served by the same DPU, the training sequence for joining lines is designed such that active lines are not disturbed. To avoid crosstalk from joining lines into active lines, crosstalk coming from the joining lines is canceled before the joining lines start to transmit regular training signals. This is done in a two-step approach.

Assuming that there is a group $\mathbb{I}_{\text{a}}$ of active lines and another group $\mathbb{I}_{\text{j}}$ of joining lines, the crosstalk between the active lines is canceled before line joining, $\boldsymbol{G}_{\text{aa}}^{(k)}\boldsymbol{H}_{\text{aa}}^{(k)}\boldsymbol{P}_{\text{aa}}^{(k),\prime} = \boldsymbol{I}$, by the precoder matrices $\boldsymbol{P}_{\text{aa}}^{(k),\prime}$ which are extended to a larger matrix $\boldsymbol{P}^{(k)}$ during the joining. The precoder matrix $\boldsymbol{P}^{(k)}$ and the channel estimation matrix $\boldsymbol{H}_{\text{est}}$ as introduced in Sect. 3.2.3 are partitioned according to

$$\boldsymbol{P} = \begin{bmatrix} \boldsymbol{P}_{\text{aa}}^{(k)} & \boldsymbol{P}_{\text{aj}}^{(k)} \\ \boldsymbol{P}_{\text{ja}}^{(k)} & \boldsymbol{P}_{\text{jj}}^{(k)} \end{bmatrix}, \quad \boldsymbol{H}_{\text{est}} = \begin{bmatrix} \boldsymbol{H}_{\text{est,aa}}^{(k)} & \boldsymbol{H}_{\text{est,aj}}^{(k)} \\ \boldsymbol{H}_{\text{est,ja}}^{(k)} & \boldsymbol{H}_{\text{est,jj}}^{(k)} \end{bmatrix}, \tag{3.43}$$

for joining and active lines. The equalizer diagonal matrices of the active lines are $\boldsymbol{G}_{\text{aa}}^{(k)}$ for each carrier k. The downstream precoder training is done in two steps, using two channel estimations. In the first phase, the VECTOR-1 phase [5], the feedback channel is only available on the active lines. The error feedback $e_l^{(k),[t]}$ is known from lines $l \in \mathbb{I}_{\text{a}}$ while sync symbols are transmitted on active and joining lines such that the transmit signal $u_l^{(k),[t]}$ is known for lines $l \in \mathbb{I}_{\text{a}} \cup \mathbb{I}_{\text{j}}$. The joining lines do not affect the active lines as long as only sync symbols are transmitted, because these symbols are reserved for channel estimation. With that, channel estimation is performed between the active lines, which gives $\boldsymbol{H}_{\text{est,aa}}^{(k)}$, and from the joining lines into the active lines, which gives $\boldsymbol{H}_{\text{est,aj}}^{(k)}$. With a known channel estimation matrix $\boldsymbol{H}_{\text{est,v1}}^{(k)}$ according to

[8] Accordingly, the channel estimation with the received signal vector $\hat{\boldsymbol{u}}^{(k),[t]}$ is given by $\boldsymbol{H}_{\text{est}}^{(k)} = \frac{1}{T}\sum_{t=1}^{T} \hat{\boldsymbol{u}}^{(k)[t]}\boldsymbol{u}^{(k)[t],\text{H}}$.

$$\boldsymbol{H}_{\text{est,v1}}^{(k)} \approx \boldsymbol{G}_{\text{aa}}^{(k)} \left[\boldsymbol{H}_{\text{aa}}^{(k)} \; \boldsymbol{H}_{\text{aj}}^{(k)} \right] \begin{bmatrix} \boldsymbol{P}_{\text{aa}}^{(k),'} & \boldsymbol{0} \\ \boldsymbol{0} & \boldsymbol{I} \end{bmatrix}, \tag{3.44}$$

crosstalk between active lines and crosstalk from joining into active lines can be canceled, using the method described in Appendix B.7. After the VECTOR-1 phase, joining lines can transmit signals in downstream without disturbing the active lines.

In the second phase, VECTOR-2 [5], the feedback channel for the joining lines is established, allowing a complete channel estimation between joining and active lines. With that, the remaining parts of the precoder matrix $\boldsymbol{P}_{\text{ja}}^{(k)}$ and $\boldsymbol{P}_{\text{jj}}^{(k)}$ are calculated. More details on the joining procedure can be found in Appendices B.7 and B.8.

After the joining phase, all active lines continue to update the precoder and equalizer matrices to follow changes of the channel and noise conditions.

3.2.5 Linear Zero-Forcing Precoding

Zero-forcing precoding is a common approach for VDSL2 precoding as defined in the G.vector standard [8]. As explained in [21] it is close to the optimum solution due to the structure of the channel matrices for copper cable bundles at lower frequencies. G.fast precoding operates at much higher frequencies, which requires precoding improvements compared to G.vector. Nevertheless, precoding is based on the zero-forcing condition

$$\boldsymbol{G}^{(k)} \boldsymbol{H}^{(k)} \boldsymbol{P}^{(k)} = \boldsymbol{I}, \tag{3.45}$$

where $\boldsymbol{I}$ is the $L \times L$ identity matrix.

The precoder matrix $\boldsymbol{P}^{(k)}$ is the scaled inverse of the channel matrix

$$\boldsymbol{P}^{(k)} = \boldsymbol{H}^{(k),-1} \text{diag}\left(\text{diag}\left(\boldsymbol{H}^{(k),-1}\right)\right)^{-1}. \tag{3.46}$$

Without loss of generality, scaling to unit diagonal elements is assumed in Eq. (3.46) for the matrix $\boldsymbol{P}^{(k)}$ while any other scaling is performed with the diagonal gain matrix $\boldsymbol{S}^{(k)}$ according to Eq. (3.35).

The equalizer matrix $\boldsymbol{G}^{(k)}$ is given by

$$\boldsymbol{G}^{(k)} = \text{diag}\left(\text{diag}\left(\boldsymbol{H}^{(k)} \boldsymbol{P}^{(k)}\right)\right)^{-1}. \tag{3.47}$$

It must be noted that the equalizer is not under control of the DPU. CPEs may use a different optimization criteria, e.g., minimum mean squared error, to derive the equalizer coefficients.

In some cases, especially at high frequencies, only a subset $\mathbb{I}_{\text{a}}^{(k)} \subseteq \{1, \ldots, L\}$ of all lines can use a certain carrier k for data transmission, while for the remaining lines, the channel quality on carrier k is insufficient to transmit at least one bit per symbol. In this case, crosstalk cancelation is only required between the active lines, while transmitters corresponding to the inactive lines are still present, which allows a non-square precoder $\boldsymbol{P}^{(k)} \in \mathbb{C}^{L \times L_{\mathrm{a}}^{(k)}}$ where $L_{\mathrm{a}}^{(k)} = \left|\mathbb{I}_{\mathrm{a}}^{(k)}\right|$. In this case, the precoder matrix is the pseudoinverse of the reduced channel $\boldsymbol{H}^{(k)} \in \mathbb{C}^{L_{\mathrm{a}}^{(k)} \times L}$ according to

$$\boldsymbol{P}^{(k)} = [\boldsymbol{H}^{(k)}]^{+} \cdot \mathrm{diag}(\mathrm{diag}([\boldsymbol{H}^{(k)}]^{+}))^{-1} \tag{3.48}$$

where $[]^{+}$ denotes the Moore–Penrose pseudoinverse. Instead of the Moore–Penrose pseudoinverse, an optimized inverse as described in Appendix C.4 can be used, which requires an additional optimization step.

Direct channel scaling is implemented with the scale matrix $\boldsymbol{S}^{(k)}$ which is known to the receivers as proposed by the G.fast standard [5]. In contrast to VDSL2 vectoring, the transmit power increase caused by precoding in downstream direction is not negligible and several methods have been proposed to guarantee that the transmit power constraints are satisfied. They are presented in Sects. 3.2.6 and 3.2.7.

3.2.6 Column Norm Scaling and Column Reduction

The task of spectrum shaping is of increasing importance for G.fast, compared to Vectored VDSL2. While for VDSL2, spectrum optimization is performed as described in Sect. 3.1.6, while the influence of the precoder on the spectrum is negligible, this is not possible for G.fast due to the increased crosstalk level.

Column norm scaling is proposed in [26] as a low complexity method to satisfy the spectral constraints. It is used as a reference for the spectrum optimization methods discussed in Sects. 3.2.8 and 3.2.9.

The column norm scaling method as proposed in [26] handles only the per-line spectral mask constraint (Eq. (3.37)), but not the per-line sum-power constraint (Eq. (3.38)). Therefore, the per-line sum-power constraint is incorporated into a new spectral mask

$$\hat{p}_{\mathrm{mask}}^{(k)} = \min(p_{\mathrm{mask}}^{(k)}, \mu_{\mathrm{sum}}) \tag{3.49}$$

to satisfy both, the per-line sum-power constraint and the per-line spectral mask constraint with column norm scaling. This is done by a flat PSD cut-back μ_{sum} which is selected to satisfy

$$\max \mu_{\mathrm{sum}} \text{ s.t. } \sum_{k=1}^{K} \hat{p}_{\mathrm{mask}}^{(k)} \leq p_{\mathrm{sum}}. \tag{3.50}$$

The power per carrier $x_l^{(k)}$ is calculated in a second step with respect to the precoder matrices. Therefore, the inverse of the column norm of the inverse channel matrix

$$\boldsymbol{S}_{\mathrm{c}}^{(k)} = \sqrt{\mathrm{diag}\left(\mathrm{diag}\left([\boldsymbol{H}^{(k)}]^{+,\mathrm{H}}[\boldsymbol{H}^{(k)}]^{+}\right)\right)}^{-1} \tag{3.51}$$

Algorithm 2 Spectrum shaping using column norm scaling [26]

Reduce transmit PSD to satisfy per-line sum-power constraint, Eqs. (3.49) and (3.50)
Initialize $\mathbb{I}_{\mathrm{a}}^{(k)} = \{1, \ldots, L\}$
repeat
  Calculate channel pseudoinverse $\left[\boldsymbol{H}^{(k)}\right]^{+}$ for given $\mathbb{I}_{\mathrm{a}}^{(k)}$ with Eq. (3.48)
  Calculate column norm $\boldsymbol{S}_{\mathrm{c}}^{(k)}$ with Eq. (3.51)
  Check the column norm values to be above s_{th}
  Update $\mathbb{I}_{\mathrm{a}}^{(k)}$
until All scaling values are above lower bound
Calculate power allocation Eq. (3.52)

is calculated [26].[9] The transmit power value is selected for all lines such that the PSD limit is satisfied

$$x_l^{(k)} = \frac{\hat{p}_{\mathrm{mask}}^{(k)} \left[\boldsymbol{S}_{\mathrm{c}}^{(k),2}\right]_{ll}}{\max\left(\operatorname{diag}\left(\boldsymbol{P}^{(k)}\boldsymbol{S}_{\mathrm{c}}^{(k)}\boldsymbol{S}_{\mathrm{c}}^{(k),\mathrm{H}}\boldsymbol{P}^{(k),\mathrm{H}}\right)\right)}. \tag{3.52}$$

Besides that, matrix inversion according to Eq. (3.46) may be ill conditioned and the normalization according to (3.52) will give very small values of $x_l^{(k)}$. In this case, the set of active lines $\mathbb{I}_{\mathrm{a}}^{(k)}$ on the specific carrier is reduced and the precoder is recomputed according to Eq. (3.48). Equation (3.52) is repeated for the reduced channel matrix. A threshold s_{th} on the column norm is used to preselect the active channels, e.g., $\mathbb{I}_{\mathrm{a}}^{(k)} = \left\{i : \left[\boldsymbol{S}_{\mathrm{c}}^{(k)}\right]_{\mathrm{ii}} \geq s_{\mathrm{th}}\right\}$.

Algorithm 2 summarizes the column norm scaling approach. This method can be implemented with low computational effort and avoids that the power constraints are violated in precoded G.fast, but it is not optimal in terms of data rate.

3.2.7 Greedy Bit Allocation Methods

Besides column norm scaling, there is another class of spectrum allocation algorithms to be considered for comparison with the proposed spectrum optimization in Sect. 3.2.8. These methods directly optimize the discrete bit allocation function according to Eq. (3.16). They are based on a greedy approach, where bit and power allocation are derived such that one bit is assigned where it consumes the lowest power and does not cause power constraint violations. Such methods have been introduced for single DSL lines [28] and extended for linear precoding [29] and nonlinear precoding [15].

The algorithm of [29] stops at some sub-optimal power allocation whenever it is not possible to increase the transmit power such that one more bit can be allocated at any carrier, which is not necessarily optimal. The method of [15] improves this

[9] Another approach is proposed in [27] and uses $\boldsymbol{S}_{\mathrm{c}}^{(k)} = \operatorname{diag}\left(\boldsymbol{H}^{(k)}\right)$.

behavior by a two-step approach, where in the first step, only the spectral mask constraints are considered. In a second step, the bit allocation is reduced until the per-line sum-power constraint is met. The simulations presented in Appendix B.17 show that the greedy bit allocation algorithm does not achieve results close to the optimal performance.

Reference [30] describes a spectrum allocation algorithm to solve the discrete bit allocation optimally. However, the combinatorial search used in the algorithm is not feasible in practice due to the high complexity.

3.2.8 Spectrum Optimization via Convex Optimization

While the previously discussed methods in Sects. 3.2.6 and 3.2.7 are not based on convex optimization methods, the following sections discuss methods to perform precoding and spectrum optimization based on convex optimization. The presented algorithms focus on optimal solutions for maximization of the sum-rate $\sum_{l=1}^{L} R_l$ for a given zero-forcing precoder and are designed such that an efficient implementation for G.fast is possible.

Discrete Rate Objective and Continuous Approximations

The data rate in G.fast is a discrete function, as described in Sect. 3.1.5. To apply convex optimization methods for G.fast rate optimization, a concave and continuous approximation of the rate objective is required.

The data rate can be approximated by the term $\log_2\left(1+\frac{SNR^{(k)}}{\Gamma}\right)$ [12], but the approximation does not take into account that the data rate is lower bounded by the smallest 1 bit constellation and upper bounded by the constellation with $b_{\max}$ bits.

Following the approach of Sect. 3.1.6 convex rate optimization is based the rate function with upper and lower bound considered according to

$$b^{(k)} = \begin{cases} 0 & SNR^{(k)} \leq SNR_{\min} \\ \log_2\left(1+\frac{SNR^{(k)}}{\Gamma}\right) & SNR_{\min} < SNR^{(k)} < SNR_{\max} \\ b_{\max} & SNR^{(k)} \geq SNR_{\max} \end{cases}. \tag{3.53}$$

Equation (3.53) is concave within $SNR_{\min} < SNR < SNR_{\max}$. For zero-forcing precoding, the signal-to-noise ratio is given by

$$SNR_l^{(k)} = \frac{x_l^{(k)}}{|g_l^{(k)}|^2\sigma^2} \tag{3.54}$$

with the receive equalizer for line l and carrier k to be $g_l^{(k)} = \frac{1}{\boldsymbol{h}_l^{(k),\mathrm{T}}\boldsymbol{p}_l^{(k)}}$. The channel matrix is partitioned into one row vector $\boldsymbol{h}_l^{(k),\mathrm{T}}$ per line and the precoder matrix is divided into one column vector $\boldsymbol{p}_l^{(k)}$ per line. Equation (3.54) linear in $\boldsymbol{x}^{(k)}$ and the SNR bounds can be translated into the upper and lower bound of the transmit power, which will be discussed in the next section.

Power Constraints for Convex Spectrum Optimization

The transmit spectrum is controlled by the diagonal gain matrix $\mathbf{S}^{(k)}$ at the precoder input and the power constraints according to Eqs. (3.37) and (3.38) are formulated with respect to $\mathbf{S}^{(k)}$. For the presented spectrum optimization algorithms, it is more convenient to optimize for the transmit power $x_l^{(k)} = |s_l^{(k)}|^2$ at the precoder input.

The per-carrier constraints are formulated as linear inequalities of the form $\boldsymbol{A}^{(k)}\boldsymbol{x}^{(k)} \leq \boldsymbol{d}^{(k)}$ with the constraint matrix $\boldsymbol{A}^{(k)}$ and the constraint vector $\boldsymbol{d}^{(k)}$. The constraint set consists of the spectral mask constraint to be $\boldsymbol{P}^{(k)} \odot \boldsymbol{P}^{(k),*}\boldsymbol{x}^{(k)} \leq \boldsymbol{p}_{\text{mask}}^{(k)}$, the positiveness constraint on $\boldsymbol{x}^{(k)}$ and the power constraint according to Eq. (3.27) to take the bit loading upper bound into account. For linear zero-forcing precoding, the upper bound $\boldsymbol{p}_{\text{bmax}}^{(k)} = [p_{\text{bmax},1}^{(k)}, \ldots, p_{\text{bmax},L}^{(k)}]^{\text{T}}$ is given by

$$p_{\text{bmax},l}^{(k)} = \left(2^{b_{\max}} - 1\right) |g_l^{(k)}|^2 \Gamma \sigma^2. \tag{3.55}$$

The constraint matrix $\boldsymbol{A}^{(k)}$ and the constraint vector $\boldsymbol{d}^{(k)}$ are given by

$$\boldsymbol{A}^{(k)} = \begin{pmatrix} \boldsymbol{P}^{(k)} \odot \boldsymbol{P}^{(k),*} \\ \boldsymbol{I}_L \\ -\boldsymbol{I}_L \end{pmatrix} \quad \boldsymbol{d}^{(k)} = \begin{pmatrix} \boldsymbol{p}_{\text{mask}}^{(k)} \\ \boldsymbol{p}_{\text{bmax}}^{(k)} \\ \boldsymbol{0}_L \end{pmatrix}. \tag{3.56}$$

To satisfy the per-line sum-power constraint, the reduced spectral mask $\hat{p}_{\text{mask}}^{(k)}$ according to Eq. (3.49) is used instead of the limit mask $p_{\text{mask}}^{(k)}$ in Eq. (3.56) or an additional constraint of the form

$$\sum_{k=1}^{K} \boldsymbol{A}_{\text{sum}}^{(k)} \boldsymbol{x}^{(k)} \leq \boldsymbol{p}_{\text{sum}} \tag{3.57}$$

with is $\boldsymbol{A}_{\text{sum}}^{(k)} = \boldsymbol{P}^{(k)} \odot \boldsymbol{P}^{(k),*}$ is required.

The transmit power can be maximized by solving the linear program

$$\min_{\boldsymbol{x}^{(k)}} \boldsymbol{c}^{(k),T} \boldsymbol{x}^{(k)} \text{ s. t. } \boldsymbol{A}^{(k)} \boldsymbol{x}^{(k)} \leq \boldsymbol{d}^{(k)} \tag{3.58}$$

for each carrier k with the objective vector $\boldsymbol{c}^{(k)}$ according to

$$c_l^{(k)} = \boldsymbol{p}_l^{(k),\text{H}} \boldsymbol{p}_l^{(k)}. \tag{3.59}$$

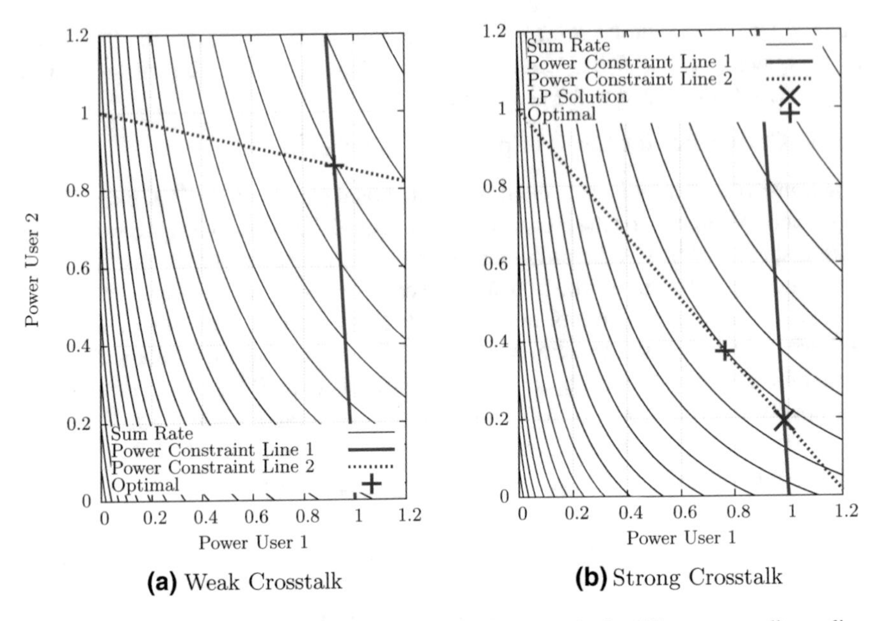

(a) Weak Crosstalk **(b)** Strong Crosstalk

Fig. 3.12 Optimal power allocation for linear zero-forcing precoder for different crosstalk coupling strength.eps

This gives higher data rates than column norm scaling, but it is not always optimal in terms of data rate, as Fig. 3.12 shows. The solution of the linear program is always on an intersection of constraints, as Fig. 3.12 demonstrates for two lines and one carrier. The contour lines indicate the sum rate of the two lines. The blue lines illustrate the per-line power constraints for both lines. For a low crosstalk channel, the linear program solution matches the rate-optimal solution (see Fig. 3.12a). For a high crosstalk channel, as shown in Fig. 3.12b and the actual sum-rate objective must be used to maximize the data rate.

The optimal power vector $\boldsymbol{x}^{(k)}$ may have zero elements, which means that crosstalk into such lines is not required to be canceled because the corresponding carrier and line does not have any power allocated to it. This results in additional degrees of freedom for the matrix inversion.

The work of [31] shows methods to use these degrees of freedom and find the optimal pseudoinverse. A simplified approach is based on the pseudoinverse as defined in Eq. (3.48). A generalized matrix inversion that fits to the G.fast spectrum optimization problem is described in Appendix C.4.

Algorithm 3 summarizes linear programming-based spectrum optimization with carrier deactivation. Hereby, the set of active lines $\mathbb{I}_{\mathrm{a}}^{(k)}$ on each carrier is reduced successively until all lines achieve the minimum bit loading, e.g., $b_{\min} = 1$.

Algorithm 3 Optimized Linear Precoding using Linear Programming

Derive reduced spectral mask $\hat{p}_{\text{mask}}^{(k)}$ (Eq. (3.49))
Start with initial set of active lines $\mathbb{I}_{\text{a}}^{(k)} = \{1, \ldots, L\}$
Initialize precoder matrices $\boldsymbol{P}^{(k)}$
Derive constraint set according to Eq. (3.55) and (3.56)
Calculate objective (3.59)
repeat
  Solve Eq. (3.58)
  if $\exists x_l^{(k)} = 0 : l \in \mathbb{I}_{\text{a}}^{(k)}$ **then**
    Remove the weakest link $\arg\min_{l \in \mathbb{I}_{\text{a}}} \left| \boldsymbol{h}_l^{(k),\text{T}} \boldsymbol{p}_l^{(k)} \right|^2$ from the set of active lines
    Re-compute Eqs. (3.55), (3.56), (3.59)
  end if
until $x_l^{(k)} > 0 \,\forall\, l \in \mathbb{I}_{\text{a}}^{(k)}$

A similar approach, ignoring the bit loading lower bound and the bit loading upper bound is reported in [32]. The method has been published earlier in [33].

Successive Quadratic Optimization

For a given zero-forcing precoder, the convex spectrum optimization problem with linear power constraints can be solved globally optimal. This section presents the problem formulation as well as a spectrum optimization algorithm for G.fast. The following algorithm is published in [2]. The optimization objective is the sum of bits according to Eq. (3.53) which gives

$$R_l = \sum_{k=1}^{K} \log_2 \left(1 + \frac{x_l^{(k)}}{\Gamma \sigma^2 |g_l^{(k)}|^2} \right). \tag{3.60}$$

The spectrum optimization problem for linear zero-forcing precoding with per-line sum-power and spectral mask constraints reads as

$$\begin{aligned} \min_{\boldsymbol{x}^{(1)},\ldots,\boldsymbol{x}^{(K)}} \sum_{l=1}^{L} -R_l \quad & \text{s.t. } \boldsymbol{A}^{(k)} \boldsymbol{x}^{(k)} \leq \boldsymbol{d}^{(k)} \,\forall\, k = 1, \ldots, K \\ & \text{s.t. } \sum_{k=1}^{K} \boldsymbol{A}_{\text{sum}}^{(k)} \boldsymbol{x}^{(k)} \leq \boldsymbol{p}_{\text{sum}}, \end{aligned} \tag{3.61}$$

where the constraint matrix $\boldsymbol{A}^{(k)}$ and the constraint vector $\boldsymbol{d}^{(k)}$ are given by Eq. (3.56). The sum-power constraint matrix is $\boldsymbol{A}_{\text{sum}}^{(k)} = \boldsymbol{P}^{(k)} \odot \boldsymbol{P}^{(k),*}$.

Equation (3.61) is a convex optimization problem because the constraints are linear inequalities and $\sum_{l=1}^{L} -R_l$ is a convex function.

The optimization problem of Eq. (3.61) is coupled over all lines and all subcarriers through the per-line sum-power constraint. This is a disadvantage in terms of complexity, because the number of subcarriers K is high, e.g., $K = 2048$ or $K = 4096$ for G.fast and the typical number of lines is around $L = 16$ or $L = 24$, up to $L = 128$.

Lagrange duality allows to separate the problem into a per-carrier subproblem and a sum-power allocation problem.[10] The Lagrangian $\phi(\boldsymbol{x}^{(k)}, \boldsymbol{\mu}_{\text{sum}})$ of the per-carrier subproblem

$$\phi(\boldsymbol{x}^{(k)}, \boldsymbol{\mu}_{\text{sum}}) = \sum_{l=1}^{L} -R_l + \boldsymbol{\mu}_{\text{sum}}^{\text{T}} \left(\sum_{k=1}^{K} \boldsymbol{A}_{\text{sum}}^{(k)} \boldsymbol{x}^{(k)} - \boldsymbol{p}_{\text{sum}} \right), \tag{3.62}$$

is convex in $\boldsymbol{x}^{(k)}$ and it is twice differentiable. The first derivative is given by

$$\frac{\partial \phi(\boldsymbol{x}^{(k)}, \boldsymbol{\mu}_{\text{sum}})}{\partial x_l^{(k)}} = -\frac{\frac{1}{\ln 2}}{x_l^{(k)} + \Gamma \sigma^2 |g_l^{(k)}|^2} + \boldsymbol{\mu}_{\text{sum}}^{\text{T}} \boldsymbol{a}_{\text{sum},l}^{(k)}, \tag{3.63}$$

where $\boldsymbol{a}_{\text{sum},l}^{(k)}$ is the lth column of $\boldsymbol{A}_{\text{sum}}^{(k)}$ and the second derivative is obtained as

$$\frac{\partial^2 \phi(\boldsymbol{x}^{(k)}, \boldsymbol{\mu}_{\text{sum}})}{\partial x_v^{(k)} \partial x_d^{(k)}} = \begin{cases} \dfrac{1}{\ln 2 \left(\Gamma \sigma^2 |g_l^{(k)}|^2 + x_v^{(k)} \right)^2} & \text{for } v = d \\ 0 & \text{for } v \neq d. \end{cases} \tag{3.64}$$

The vector $\nabla \phi(\boldsymbol{x}^{(k)}, \boldsymbol{\mu}_{\text{sum}}) = \left[\frac{\partial \phi(\boldsymbol{x}^{(k)}, \boldsymbol{\mu}_{\text{sum}})}{\partial x_1^{(k)}}, \ldots, \frac{\partial \phi(\boldsymbol{x}^{(k)}, \boldsymbol{\mu}_{\text{sum}})}{\partial x_L^{(k)}} \right]^{\text{T}}$ and the matrix $\nabla^2 \phi(\boldsymbol{x}^{(k)}, \boldsymbol{\mu}_{\text{sum}}) = \text{diag}\left(\left[\frac{\partial^2 \phi(\boldsymbol{x}^{(k)}, \boldsymbol{\mu}_{\text{sum}})}{\partial x_1^{(k)} \partial x_1^{(k)}}, \ldots, \frac{\partial^2 \phi(\boldsymbol{x}^{(k)}, \boldsymbol{\mu}_{\text{sum}})}{\partial x_L^{(k)} \partial x_L^{(k)}} \right] \right)$ can be used to approximate the objective function for one carrier k around $\boldsymbol{x}_0^{(k)}$ by

$$\begin{aligned} \phi(\boldsymbol{x}^{(k)}, \boldsymbol{\mu}_{\text{sum}}) \approx\ & \phi(\boldsymbol{x}_0^{(k)}, \boldsymbol{\mu}_{\text{sum}}) + (\boldsymbol{x}^{(k)} - \boldsymbol{x}_0^{(k)})^{\text{T}} \nabla \phi(\boldsymbol{x}^{(k)}, \boldsymbol{\mu}_{\text{sum}}) \\ & + (\boldsymbol{x}^{(k)} - \boldsymbol{x}_0^{(k)})^{\text{T}} \nabla^2 \phi(\boldsymbol{x}^{(k)}, \boldsymbol{\mu}_{\text{sum}}) (\boldsymbol{x}^{(k)} - \boldsymbol{x}_0^{(k)}). \end{aligned} \tag{3.65}$$

It is proposed to solve for the optimal power allocation $\boldsymbol{x}^{(k)}$, using Newton steps. The objective Eq. (3.65) is brought to the form $f_{\text{q}}(\boldsymbol{x}^{(k)}, \boldsymbol{x}_0^{(k)}) = \boldsymbol{c}^{\text{T}} \boldsymbol{x}^{(k)} + \frac{1}{2} \boldsymbol{x}^{(k),\text{T}} \boldsymbol{H}_{\text{q}} \boldsymbol{x}^{(k),\text{T}}$ with $\boldsymbol{c}_q = \nabla \Phi(\boldsymbol{x}^{(k)}, \boldsymbol{\mu}_{\text{sum}}) - \nabla^2 \Phi(\boldsymbol{x}^{(k)}, \boldsymbol{\mu}_{\text{sum}}) \boldsymbol{x}_0^{(k)}$ and $\boldsymbol{H}_q = \nabla^2 \Phi(\boldsymbol{x}^{(k)}, \boldsymbol{\mu}_{\text{sum}})$ while the constant terms are neglected.

In each step

$$\boldsymbol{x}^{(k),[t]} = \arg \min_{\boldsymbol{x}^{(k),[t]}} f_q(\boldsymbol{x}^{(k),[t]}, \boldsymbol{x}^{(k),[t-1]}), \ \text{s.t.}\ \boldsymbol{A}^{(k)} \boldsymbol{x}^{(k),[t]} \leq \boldsymbol{d}^{(k)} \tag{3.66}$$

The solution of Eq. (3.66) is used as a starting point for the next iteration. A small number of quadratic approximation steps is required for convergence.

[10] Equation (3.49) in Sect. 3.2.6 shows an alternative suboptimal method to decouple the spectrum optimization problem over the subcarriers.

The sum-power allocation problem is solved using a projected gradient approach. The gradient projection step for $\boldsymbol{\mu}_{\text{sum}}$ is

$$\boldsymbol{\mu}_{\text{sum}}^{[t+1]} = \max\left(\boldsymbol{\mu}_{\text{sum}}^{[t]} + \alpha_{\text{sum}}\left(\sum_{k=1}^{K} \boldsymbol{A}_{\text{sum}}^{(k)} \boldsymbol{x}^{(k)} - \boldsymbol{p}_{\text{sum}}\right), 0\right) \tag{3.67}$$

with the step size α_{sum}.

The proposed method is summarized in Algorithm 9 in Appendix B.9. It represents a low complexity, but globally optimal solution for the spectrum optimization problem for liner zero-forcing precoding which meets the requirements of a G.fast system, as the implementation-specific constraints are considered. A lower complexity approximation of Algorithm 9 is presented in Appendix B.9.

While Algorithms 3 and 9 rely on the zero-forcing condition, there is an alternative approach to optimize precoder and transmit spectrum without zero-forcing constraint.

3.2.9 Linear MMSE Precoding via Uplink/Downlink Duality

In contrast to zero-forcing precoding with spectrum optimization, minimum mean squared error (MMSE) precoding allows joint optimization of precoder coefficients and spectrum. The zero-forcing constraint, Eq. (3.45), is removed and interference is allowed. This increases the degrees of freedom for spectrum optimization and results in a potential rate increase compared to the method discussed in Sect. 3.2.8.

This section shows how to apply known results on uplink-downlink duality with per-antenna power constraints [34], MSE minimization with per-antenna power constraints [35] and weighted sum-rate maximization by weighted MSE minimization [36] on G.fast precoder optimization.

Despite the availability of these results, MMSE precoding for G.fast requires to consider additional conditions, which are

- rate optimization with an SNR gap $\Gamma > 1$,
- the bit loading upper bound $b_l^{(k)} \leq b_{\max}$ and
- power constraints in the form of per-line sum-power and per-line spectral mask constraints.

Available results on MSE minimization with SNR gap [37] and MSE minimization with multiple power constraints [38] are considered to derive the MSE precoding algorithm for G.fast, which is based on uplink-downlink duality [39]. A computationally efficient implementation for MMSE precoding, based on the primal problem, is presented in Sect. 3.2.10 and the channel capacity for linear precoding for G.fast is presented in Sect. 3.5. The MMSE precoding approach presented in this section is published in [2].

MSE Minimization with SNR Gap

Sum-rate maximization can be solved by multiple weighted MSE minimizations, with properly selected weights as shown in [36]. The receiver error vector is $\boldsymbol{e}^{(k)} = \hat{\boldsymbol{u}}^{(k)} - \boldsymbol{u}^{(k)}$ The per-line MSE $\xi_l^{(k)}$ for line l is given by

$$\begin{aligned}\xi_l^{(k)} = \mathrm{E}\left[|e_l^{(k)}|^2\right] &= |g_l^{(k)}|^2\sigma^2 + |g_l^{(k)}|^2\boldsymbol{h}_l^{(k),\mathrm{T}}\boldsymbol{p}_l^{(k)}\boldsymbol{p}_l^{(k),\mathrm{H}}\boldsymbol{h}_l^{(k),*} \\ &- 2\mathrm{Re}\{g_l^{(k)}\boldsymbol{h}_l^{(k),\mathrm{T}}\boldsymbol{p}_l^{(k)}\} + \sum_{d\neq l}|g_l^{(k)}\boldsymbol{h}_l^{(k),\mathrm{T}}\boldsymbol{p}_d^{(k)}|^2 + 1\end{aligned} \tag{3.68}$$

where $\boldsymbol{\xi}^{(k)}$ is the vector of MSEs of all lines for carrier k.

For zero-forcing precoding, capacity is given by $\log_2\left(1 + SNR^{(k)}\right)$ and the data rate with SNR gap, is approximated by $\log_2\left(1 + \frac{SNR^{(k)}}{\Gamma}\right)$. For MMSE precoding, assuming that the receiver uses the MSE optimal equalizer $\boldsymbol{G}^{(k),\star}$, capacity is given by $\log_2\left(\frac{1}{\xi_l^{(k)}}\right)$. Introducing the SNR gap in the minimum MSE solution, an equivalent MSE $\xi_{\Gamma,l}^{(k)}$ is introduced such that the rate can be expressed as $\log_2\left(\frac{1}{\xi_{\Gamma,l}^{(k)}}\right)$.

Consequently, $1 + \frac{SNR}{\Gamma} = \frac{1}{\xi_\Gamma}$ must be satisfied, which leads to

$$1 + \frac{SNR_l^{(k)}}{\Gamma} = \frac{\left|\boldsymbol{h}_l^{(k),\mathrm{T}}\boldsymbol{p}_l^{(k)}\right|^2 + \Gamma\sum_{d\neq l}\left|\boldsymbol{h}_l^{(k),\mathrm{T}}\boldsymbol{p}_d^{(k)}\right|^2 + \Gamma\sigma^2}{\Gamma\sum_{d\neq l}\left|\boldsymbol{h}_l^{(k),\mathrm{T}}\boldsymbol{p}_d^{(k)}\right|^2 + \Gamma\sigma^2} \tag{3.69}$$

and indicates that the equivalent MSE $\boldsymbol{\xi}_\Gamma^{(k)}$ to be optimized in presence of the SNR gap is

$$\xi_{\Gamma,l}^{(k)} = 1 - \frac{|\boldsymbol{h}_l^{(k),\mathrm{T}}\boldsymbol{p}_l^{(k)}|^2}{|\boldsymbol{h}_l^{(k),\mathrm{T}}\boldsymbol{p}_l^{(k)}|^2 + \Gamma\sum_{d\neq l}|\boldsymbol{h}_l^{(k),\mathrm{T}}\boldsymbol{p}_d^{(k)}|^2 + \Gamma\sigma^2} \tag{3.70}$$

for the optimal receive equalizer $\boldsymbol{G}_\Gamma^{(k),\star}$. The optimal receiver with an SNR gap unequal 1 is given by

$$g_{\Gamma,ll}^{(k),\star} = \frac{\left(\boldsymbol{h}_l^{(k),\mathrm{T}}\boldsymbol{p}_l^{(k)}\right)^*}{|\boldsymbol{h}_l^{(k),\mathrm{T}}\boldsymbol{p}_l^{(k)}|^2 + \Gamma\left(\sum_{d\neq l}|\boldsymbol{h}_l^{(k),\mathrm{T}}\boldsymbol{p}_d^{(k)}|^2 + \sigma^2\right)}. \tag{3.71}$$

Weighted MSE Minimization with Bit Loading Constraint

The weighted MMSE optimization problem for given weights is

$$\min_{\substack{\boldsymbol{P}^{(1)},\ldots,\boldsymbol{P}^{(K)}\\ \boldsymbol{G}^{(1)},\ldots,\boldsymbol{G}^{(K)}}} \sum_{k=1}^{K} \boldsymbol{w}^{(k),\mathrm{T}} \boldsymbol{\xi}_{\Gamma}^{(k)} \tag{3.72}$$

$$\text{s.t. } \operatorname{diag}\left(\boldsymbol{P}^{(k)}\boldsymbol{P}^{(k),\mathrm{H}}\right) \leq \boldsymbol{p}_{\text{mask}}^{(k)} \; \forall\, k = 1, \ldots, K \tag{3.73}$$

$$\text{s.t. } \frac{1}{\xi_{\Gamma,l}^{(k)}} \geq 2^{b_{\max}} \; \forall\, k = 1, \ldots, K \tag{3.74}$$

$$\text{s.t. } \sum_{k=1}^{K} \operatorname{diag}\left(\boldsymbol{P}^{(k)}\boldsymbol{P}^{(k),\mathrm{H}}\right) \leq \boldsymbol{p}_{\text{sum}}. \tag{3.75}$$

Following [36], sum-rate optimization is performed by sequential weighted MSE minimizations with weights $\boldsymbol{w}^{(k)} = \left[w_1^{(k)}, \ldots, w_L^{(k)}\right]^{\mathrm{T}}$ which are the inverse of the MSE $\boldsymbol{\xi}_{\Gamma}$,

$$w_l^{(k)} = \frac{1}{\xi_{\Gamma\,l}^{(k)}}. \tag{3.76}$$

After multiple updates of precoder and weights, this converges to the sum-rate optimal solution.

Strong duality holds between uplink and downlink [38] for a constraint of the form

$$\sum_{l=1}^{L} \boldsymbol{p}_l^{(k),\mathrm{H}} \boldsymbol{A}_l^{(k)} \boldsymbol{p}_l^{(k)} \leq d_l^{(k)}. \tag{3.77}$$

To achieve that, the zero-forcing approximation of the maximum bit loading constraint of Eq. (3.74) is used. As shown in Appendix B.12, this is possible for the given optimization problem.

Uplink-Downlink Duality

While the optimal receive equalizers for fixed precoders are given by Eq. (3.71), optimizing the precoders for fixed equalizers is a non-convex problem. The optimal precoders can be found in the dual uplink applying uplink-downlink duality.

After calculating the equalizer using Eq. (3.71), precoder optimization is performed in the dual uplink. The dual uplink system model in comparison to downlink is shown in Fig. 3.13.

The downlink equalizers $\boldsymbol{G}^{(k)}$ are transformed into uplink precoders $\boldsymbol{P}_{\mathrm{u}}^{(k)} = \boldsymbol{G}^{(k),\mathrm{H}}$. The uplink channel is defined as $\boldsymbol{H}_{\mathrm{u}}^{(k)} = \boldsymbol{H}^{(k),\mathrm{H}}$. We define an effective dual uplink channel $\bar{\boldsymbol{H}}_{\mathrm{u}}^{(k)} = \boldsymbol{H}_{\mathrm{u}}^{(k)}\boldsymbol{P}_{\mathrm{u}}^{(k)} = [\bar{\boldsymbol{h}}_{\mathrm{u},1}^{(k)}, \ldots, \bar{\boldsymbol{h}}_{\mathrm{u},L}^{(k)}]$. The dual uplink model is used to calculate the dual uplink equalizer $\boldsymbol{G}_{\mathrm{u}}^{(k)}$. A column vector representation of

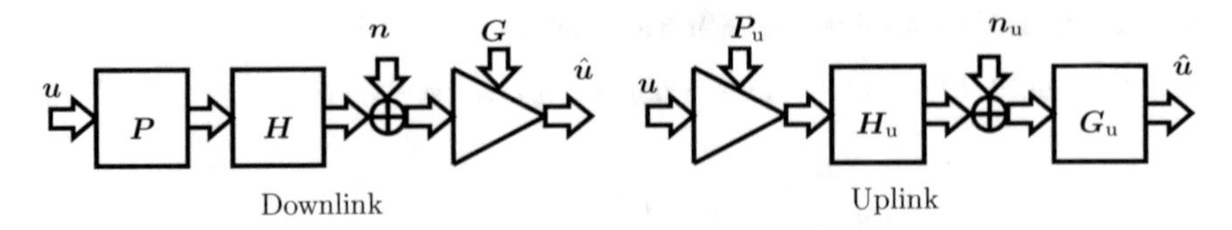

Fig. 3.13 System model for downlink precoding with the corresponding model for the dual uplink

$\boldsymbol{G}_{\mathrm{u}}^{(k)} = \left[\boldsymbol{g}_{\mathrm{u},1}, \ldots, \boldsymbol{g}_{\mathrm{u},L}\right]$ as well as a row vector representation, denoted by $\bar{\boldsymbol{g}}_{\mathrm{u},l}$ as $\boldsymbol{G}_{\mathrm{u}}^{(k)} = \left[\bar{\boldsymbol{g}}_{\mathrm{u},1}, \ldots, \bar{\boldsymbol{g}}_{\mathrm{u},L}\right]^{\mathrm{T}}$ are required for the next steps.

The dual uplink MSE for the system shown in Fig. 3.13 is given by

$$\begin{aligned}\xi_{\mathrm{u},l}^{(k)} =& \bar{\boldsymbol{g}}_{\mathrm{u},l}^{(k),\mathrm{T}} \boldsymbol{Q}^{(k)} \bar{\boldsymbol{g}}_{\mathrm{u},l}^{(k),*} + \bar{\boldsymbol{g}}_{\mathrm{u},l}^{(k),\mathrm{T}} \bar{\boldsymbol{h}}_{\mathrm{u},l}^{(k)} \bar{\boldsymbol{h}}_{\mathrm{u},l}^{(k),\mathrm{H}} \bar{\boldsymbol{g}}_{\mathrm{u},l}^{(k),*} \\ &- 2\mathrm{Re}\{\bar{\boldsymbol{g}}_{\mathrm{u},l}^{(k),\mathrm{T}} \bar{\boldsymbol{h}}_{\mathrm{u},l}^{(k)}\} + \sum_{d \neq l} |\bar{\boldsymbol{g}}_{\mathrm{u},l}^{(k),\mathrm{T}} \bar{\boldsymbol{h}}_{\mathrm{u},d}^{(k)}|^2 + 1 \end{aligned} \tag{3.78}$$

where $\boldsymbol{Q}^{(k)}$ is the uplink noise covariance matrix. For uplink-downlink duality, $\sum \xi_{\Gamma,l} = \sum \xi_{\Gamma \mathrm{u},l}$ must hold. Comparing (3.68) and (3.78), this gives $\mathrm{tr}(\boldsymbol{G}\boldsymbol{G}^{\mathrm{H}}\sigma^2) = \mathrm{tr}(\boldsymbol{G}_u \boldsymbol{Q} \boldsymbol{G}_u^{\mathrm{H}})$.

The lth row of the optimal receive equalizer $\boldsymbol{G}_{\mathrm{u}}^{(k)}, \bar{\boldsymbol{g}}_{\mathrm{u},l}^{(k)}$, which minimizes Eq. (3.78), is given by

$$\bar{\boldsymbol{g}}_{\mathrm{u},l}^{(k),\mathrm{T}} = w_l^{(k)} \bar{\boldsymbol{h}}_l^{(k),\mathrm{H}} (\boldsymbol{R}_l^{(k)} / \bar{\sigma}^{(k)} + \boldsymbol{Q}_l^{(k)} \bar{\sigma}^{(k)})^{-1} \tag{3.79}$$

with the effective noise variance $\bar{\sigma}^{2,(k)} = \mathrm{tr}\left(\sigma^2 \mathrm{diag}\left(\boldsymbol{w}^{(k)}\right) \boldsymbol{G}^{(k)} \boldsymbol{G}^{(k),\mathrm{H}}\right)$ and with the substitution $\boldsymbol{R}_l^{(k)} = \bar{\boldsymbol{h}}_l^{(k)} \bar{\boldsymbol{h}}_l^{(k),\mathrm{H}} w_l^{(k)} + \sum_{d \neq l} \Gamma \bar{\boldsymbol{h}}_d^{(k)} \bar{\boldsymbol{h}}_d^{(k),\mathrm{H}} w_d^{(k)}$,

The dual uplink receiver noise covariance matrix $\boldsymbol{Q}^{(k)}$ depends on the Lagrangian multipliers and is given by

$$\boldsymbol{Q}_l^{(k)} = \mathrm{diag}(\boldsymbol{\mu}_{\mathrm{sum}} + \boldsymbol{\mu}_{\mathrm{mask}}^{(k)}) + \sum_{l=1}^{L} \mu_{\mathrm{bmax}\,l}^{(k)} \boldsymbol{A}_{\mathrm{bmax}\,l}^{(k)}. \tag{3.80}$$

with where $\boldsymbol{A}_{\mathrm{bmax}\,l}^{(k)} = \frac{\boldsymbol{h}_l^{(k),*} \boldsymbol{h}_l^{(k),\mathrm{T}}}{(2^{b_{\max}} - 1)\Gamma}$ with $\boldsymbol{h}_l^{(k),\mathrm{T}}$ to be the lth row of the channel matrix $\boldsymbol{H}^{(k)}$.

The Lagrangian multipliers are $\boldsymbol{\mu}_{\mathrm{mask}}^{(k)}$ for the spectral mask constraint (3.73), $\boldsymbol{\mu}_{\mathrm{bmax}}^{(k)}$ for the bit loading upper bound (3.74) and $\boldsymbol{\mu}_{\mathrm{sum}}$ for the per-line sum-power constraint (3.75). The values of the per-subcarrier Lagrange multipliers are derived by fixed-point iterations [35] using

$$\boldsymbol{\mu}_{\mathrm{mask}}^{(k),[t+1]} = \max\left(\boldsymbol{\mu}_{\mathrm{mask}}^{(k)[t]} \frac{\boldsymbol{p}_{\mathrm{tmp}}^{(k)}}{\boldsymbol{p}_{\mathrm{mask}}^{(k)}}, \varepsilon_{\mathrm{mask}}\right), \tag{3.81}$$

$$\mu_{\text{bmax}\,l}^{(k),[t+1]} = \max\left(\mu_{\text{bmax}\,l}^{(k),[t]} \frac{\beta \bar{\boldsymbol{g}}_{\text{u},l}^{(k),*} \boldsymbol{A}_{\text{bmax}\,l}^{(k)} \bar{\boldsymbol{g}}_{\text{u},l}^{(k)}}{\sigma^2}, \varepsilon_{\text{bmax}}\right). \tag{3.82}$$

The downlink transmit power vector $\boldsymbol{p}_{\text{tmp}}$ is given by

$$\boldsymbol{p}_{\text{tmp}}^{(k)} = \text{diag}(\boldsymbol{G}_u^{(k),\text{H}} \boldsymbol{G}_u^{(k)})\beta \tag{3.83}$$

with the scaling factor β which is defined in (3.86) for uplink-downlink conversion. The lower bound $\varepsilon_{\text{mask}}$ is required to guarantee that the matrix inversion for the equalizer calculation in Eq. (3.79) [35] is feasible and the lower bound $\varepsilon_{\text{bmax}}$ is required to avoid that the fixed-point iterations gets stuck at zero.

A projected gradient approach may be used instead of fixed-point iterations for the calculation of the Lagrangian multipliers [38]. For the discussed problem, the gradient projection steps are

$$\boldsymbol{\mu}_{\text{mask}}^{[t+1]} = \max\left(\boldsymbol{\mu}_{\text{mask}}^{[t]} + \alpha_{\text{mask}}(\boldsymbol{p}_{\text{tmp}} - \boldsymbol{p}_{\text{mask}}), \varepsilon_{\text{mask}}\right) \tag{3.84}$$

and

$$\mu_{\text{bmax}\,l}^{[t+1]} = \max\left(\mu_{\text{bmax}\,l}^{[t]} + \alpha_{\text{bmax}}\left(\frac{\beta \boldsymbol{g}_{\text{u}\,l}^{\text{H}} \boldsymbol{A}_{\text{bmax}\,l} \boldsymbol{g}_{\text{u}\,l}}{\sigma^2} - 1\right), 0\right) \tag{3.85}$$

with the steps sizes α_{mask} for the spectral mask constraint and α_{bmax} for the bit loading upper bound.

The scale factor $\beta^{(k)}$ for uplink-downlink conversion is

$$\beta^{(k)} = \frac{\bar{\sigma}^2}{\text{trace}(\boldsymbol{G}_u^{(k)} \boldsymbol{Q}_{\text{sum}} \boldsymbol{G}_u^{(k),\text{H}})} \tag{3.86}$$

where $\boldsymbol{Q}_{\text{sum}} = \text{diag}(\boldsymbol{\mu}_{\text{sum}} + \boldsymbol{\mu}_{\text{mask}}^{(k)}) + \sum_{l=1}^{L} \mu_{\text{bmax},l}^{(k)} \boldsymbol{A}_{\text{bmax},l}^{(k)}$ and the new downlink precoder is then

$$\boldsymbol{P}^{(k)} = \boldsymbol{G}_{\text{u}}^{(k),\text{H}} \sqrt{\beta}. \tag{3.87}$$

The per-line sum-power constraint is handled separately, because it couples the subcarriers. The Lagrangian multipliers for the per-line sum-power constraints $\boldsymbol{\mu}_{\text{sum}}$ are derived using a gradient projection approach similar to the one used for ZF precoding, Eq. (3.67). For weighted MMSE precoding, the gradient step is given by

$$\boldsymbol{\mu}_{\text{sum grad}}^{[t]} = \boldsymbol{\mu}_{\text{sum}}^{[t]} + \alpha_{\mu}\left(\sum_{k=1}^{K} \text{diag}\left(\boldsymbol{P}^{(k)} \boldsymbol{P}^{(k),\text{H}}\right) - \boldsymbol{p}_{\text{sum}}\right) \tag{3.88}$$

Algorithm 4 Weighted MMSE Precoder

Initialize $\mathbf{P}^{(k)}$ with the ZF precoder and $\boldsymbol{\mu}_{\text{sum}} = \mathbf{0}_L$
Calculate $\boldsymbol{A}_{\text{bmax},l}^{(k)}$ for all k, l
repeat
  for $k = 1$ **to** K **do**
    Initialize $\boldsymbol{\mu}_{\text{bmax}}^{(k)}$ and $\boldsymbol{\mu}_{\text{mask}}^{(k)}$
    repeat
      Calculate MSE (3.70) and update weights (3.76)
      repeat
        Update equalizer matrix $\boldsymbol{G}^{(k)}$ using (3.71)
        Downlink-uplink transform $\bar{\sigma}^{2,(k)}$, $\boldsymbol{P}_{\text{u}}^{(k)}$, $\bar{\boldsymbol{H}}_{\text{u}}^{(k)}$
        repeat
          Update noise covariance $\boldsymbol{Q}^{(k)}$ and $\boldsymbol{R}_l^{(k)}$
          Update dual uplink equalizer (3.79)
          Evaluate (3.83) and (3.86)
          Update Lagrangians using (3.81) and (3.82)
        until Convergence of $\boldsymbol{\mu}_{\text{bmax}}^{(k)}$ and $\boldsymbol{\mu}_{\text{mask}}^{(k)}$
        Uplink-downlink transform (3.87)
      until Convergence precoder and equalizer
    until Convergence of MSE weights
  end for
  Perform sum-power update using Eq. (3.88)
until Convergence of sum-power

With multiple iterations of the uplink-downlink duality, the algorithm converges. A complete algorithm for weighted MMSE precoding in G.fast is shown in Algorithm 4.

The main issue of the uplink-downlink duality approach for linear MMSE precoding is the algorithmic complexity for initialization and update of the precoder. Every precoder update step in Eq. (3.79) requires a matrix inversion inversion per precoder column.

The number of iterations until convergence of uplink-downlink iterations, the per-carrier Lagrange multipliers, the sum-power constraint and the weights for sum-rate optimization is immense for a typically sized system.

Therefore, the uplink-downlink duality approach is not suited for the G.fast initialization phase where the optimization time is limited.

3.2.10 Precoding Gradient Approaches

To achieve optimal performance of the G.fast link, precoding and spectrum allocation are continuously updated. For zero-forcing precoding, two adaptive algorithms are required. This section presents adaptive precoder and spectrum optimization methods for zero-forcing precoding as well as a gradient approach for MMSE precoder optimization.

Channel Estimation during the Tracking Phase

Adaptive precoding and spectrum optimization requires a feedback signal from the CPE receivers. This is available in G.fast via the vectoring feedback channel [5]. There are dedicated symbols, the sync symbols, which are modulated with orthogonal sequences. From this symbols with known content, the difference between transmit signal and receive signal, the error vector $\boldsymbol{e}^{(k)}$ is communicated from the CPE to the DP [40, 41], as discussed in Sect. 3.2.3.

There are two options to perform channel estimation for the tracking phase. The open-loop method requires to repeat a complete channel estimation as presented in Sect. 3.2.3 from time to time on the active G.fast link and repeat the precoder calculation according to Sects. 3.2.5 and 3.2.8. The precoder is bypassed for the sync symbols to measure crosstalk. The open-loop strategy is applicable to linear and nonlinear precoding, but it comes with several disadvantages in terms of performance and computational complexity. Averaging over many symbols is required to estimate the channel with sufficient quality and the computations to update the precoder and repeat spectrum optimization require some time. Assuming that the channel is changing slowly, but continuously, the resulting precoder and spectrum is already outdated [42]. Additional memory is required to store the temporary results of channel estimation, which increases the complexity of an implementation.

With the closed-loop update, sync symbols are precoded in the same way as regular data symbols. For linear precoding, the receive error vector is given by

$$\begin{aligned}\boldsymbol{e}^{(k)} = \hat{\boldsymbol{u}}^{(k)} - \boldsymbol{u}^{(k)} &= \boldsymbol{G}^{(k)}\left(\boldsymbol{H}^{(k)}\boldsymbol{P}^{(k)}\boldsymbol{u}^{(k)} + \boldsymbol{n}^{(k)}\right) - \boldsymbol{u}^{(k)} \\ &= \boldsymbol{G}^{(k)}\boldsymbol{H}^{(k)}\boldsymbol{P}^{(k)}\boldsymbol{u}^{(k)} + \boldsymbol{G}^{(k)}\boldsymbol{n}^{(k)} - \boldsymbol{u}^{(k)}. \end{aligned} \tag{3.89}$$

The error feedback is used to run adaptive algorithms as described Sect. 3.2.10 and the following sections to update the precoder [40, 41].

Adaptive Zero-Forcing Precoder

For the zero-forcing precoder update, the mean squared error is used as an objective. For unconstrained MSE minimization, the following optimization problem is solved

$$\min_{\boldsymbol{P}^{(k)}} \mathrm{E}\left[\mathrm{trace}\left(\boldsymbol{e}^{(k)}\boldsymbol{e}^{(k),\mathrm{H}}\right)\right]. \tag{3.90}$$

The least mean squares (LMS) algorithm [43] allows to approximate the minimum mean squared error solution from Eq. (3.90) by performing small steps in the direction of minimizing the squared error

$$\min_{\boldsymbol{P}^{(k)}} \mathrm{tr}\left(\boldsymbol{e}^{(k)}\boldsymbol{e}^{(k),\mathrm{H}}\right). \tag{3.91}$$

The corresponding precoder gradient step is

$$\frac{\partial \mathrm{tr}\left(\boldsymbol{e}^{(k)}\boldsymbol{e}^{(k),\mathrm{H}}\right)}{\partial \boldsymbol{P}^{(k),*}} = \boldsymbol{H}^{(k),\mathrm{H}}\boldsymbol{G}^{(k),\mathrm{H}}\boldsymbol{e}^{(k)}\boldsymbol{u}^{(k),\mathrm{H}} \tag{3.92}$$

which converges to the zero-forcing precoder solution, as shown in Appendix B.10.

The update rule is given by

$$\boldsymbol{P}^{(k),[t+1]} = \boldsymbol{P}^{(k),[t]} - \alpha_P \boldsymbol{H}^{(k),\mathrm{H}}\boldsymbol{G}^{(k),\mathrm{H}}\boldsymbol{e}^{(k)}\boldsymbol{u}^{(k),\mathrm{H}} \tag{3.93}$$

with a step size α_P.[11] To keep the precoder coefficients in a well defined numerical range, the diagonal coefficients are kept at a fixed value by the scaling

$$\boldsymbol{P}^{(k),[t+2]} = \boldsymbol{P}^{(k),[t+1]}\mathrm{diag}\left(\mathrm{diag}\left(\boldsymbol{P}^{(k),[t+1]}\right)\right)^{-1} \tag{3.94}$$

or by excluding the diagonal elements from the update in Eq. (3.93). The corresponding derivation and proof of convergence is shown in Annex B.10.

Adaptive Spectrum Optimization for Zero-Forcing Precoding

For optimized zero-forcing precoding, not only the precoder coefficients, but also the transmit spectrum scaling requires continuous updates.

A change of the precoder matrix requires an update of the transmit gains $s_l^{(k)}$ of all lines to satisfy the power constraints and keep the optimum data rates. The LMS update for the precoder matrix according to Eq. (3.93) can be combined with a gradient-based update of the power allocation $\boldsymbol{x}^{(k)}$. The gradient step is derived from the Lagrange function as used in Sect. 3.2.8.

The optimization problem of Eq. (3.62) is solved with a projected gradient approach, using the first derivative Eq. (3.63). The gradient step

$$\begin{aligned} x_{\mathrm{grad}\ l}^{(k)[t]} &= x_l^{(k)[t-1]} + \\ &\alpha_x \left(\left(x_l^{(k)[t-1]} + \Gamma \left| \boldsymbol{h}_l^{(k),\mathrm{T}} \boldsymbol{p}_l^{(k)} \right|^{-2} \sigma^2 \right)^{-1} - \sum_{v=1}^{L} \mu_{\mathrm{sum},v} |p_{vl}^{(k)}|^2 \right) \end{aligned} \tag{3.95}$$

with a step size α_x is performed for each time instance t.

It must be noted that the update includes the term $\left|\boldsymbol{h}_l^{(k),\mathrm{T}}\boldsymbol{p}_l^{(k)}\right|^2$ which is usually not known to the transmitter as well as the noise variance σ^2. But the transmitter can request information about the current signal-to-noise ratio $SNR_l^{(k),[t]}$ from the receiver to perform the optimization and the required term $\left|\boldsymbol{h}_l^{(k),\mathrm{T}}\boldsymbol{p}_l^{(k)}\right|^{-2}\sigma^2 = \frac{x_l^{(k),[t]}}{SNR_l^{(k),[t]}}$ is derived from that.

[11] A simplified precoder update $\boldsymbol{P}^{(k),[t+1]} = \boldsymbol{P}^{(k),[t]} - \alpha_P \boldsymbol{e}^{(k)}\boldsymbol{u}^{(k),\mathrm{H}}$ is often used instead of Eq. (3.93). For a square full-rank channel matrix, it converges to the same solution.

After the gradient step, a projection step is required to satisfy the per-carrier power constraints. Each individual constraint $\boldsymbol{a}_l^{(k),\mathrm{T}}\boldsymbol{x}_{\mathrm{grad}}^{(k)} \leq d_l^{(k)}$ with $\boldsymbol{a}_l^{(k),\mathrm{T}}$ to be the lth row of $\boldsymbol{A}^{(k)} = \left[\boldsymbol{a}_1^{(k)}, \ldots, \boldsymbol{a}_{3L}^{(k)}\right]^{\mathrm{T}}$ is checked and a set $\mathbb{I}_{\mathrm{violated}}$ of row indices of violated constraints is defined. For a certain constraint $j \in \mathbb{I}_{\mathrm{violated}}$, the projection

$$\boldsymbol{x}^{(k),[t+1]} = \boldsymbol{x}_{\mathrm{grad}}^{(k),[t]} + \boldsymbol{a}_j^{(k)} \cdot (\boldsymbol{a}_j^{(k),\mathrm{T}}\boldsymbol{a}_j)^{-1}(d_j^{(k)} - [\boldsymbol{A}^{(k)}\boldsymbol{x}_{\mathrm{grad}}^{(k),[t]}]_j) \tag{3.96}$$

is done. Eq. (3.96) is performed for the index that minimizes the distance to the constraint set $\sum_{j=1}^{3L} |\max(\boldsymbol{a}_j^{\mathrm{T}}\boldsymbol{x}^{[t+1]} - d_j, 0)|^2$. This is repeated until the constraints are satisfied within a certain precision.

Algorithm 5 Linear zero-forcing precoding gradient update

Initialize $\boldsymbol{\mu}_{\mathrm{sum}} = \boldsymbol{0}_L$
repeat
  for $k = 1$ **to** K **do**
    Receive error vector $\boldsymbol{e}^{(k)}$ from CPE side
    Update precoder according to Eq. (3.93)
    Update spectrum, Eq. (3.95)
    Update constraint set, Eq. (3.56)
    repeat
      Evaluate constraint set violations $\mathbb{I}_{\mathrm{violated}}$
      Perform projection, Eq. (3.96)
    until Constrains are satisfied
  end for
  Update power allocation using (3.67)
until Per-line sum-power is converged

The sum-power constraint is handled in the same way as for successive quadratic optimization, using Eq. (3.67). Algorithm 5 summarizes the corresponding steps of a single gradient update. It is repeated for each sync symbol or after a number of sync symbols.

Adaptive Weighted MMSE Precoding

The LMS solution to the constrained MMSE precoder can be found based on the Lagrange function of the corresponding optimization problem. The weighted MSE minimization problem reads as

$$\min_{\boldsymbol{P}^{(k)}} \sum_{k=1}^{K} \mathrm{E}\left[\mathrm{tr}\left(\boldsymbol{W}^{(k)}\boldsymbol{e}^{(k)}\boldsymbol{e}^{(k),\mathrm{H}}\right)\right] \tag{3.97}$$
$$\text{s.t. } \mathrm{diag}\left(\boldsymbol{P}^{(k)}\boldsymbol{P}^{(k),\mathrm{H}}\right) \leq \boldsymbol{p}_{\mathrm{mask}}^{(k)} \forall k = 1, \ldots, K$$
$$\text{s.t. } \mathrm{tr}\left(\boldsymbol{A}_l\boldsymbol{P}^{(k)}\boldsymbol{A}_{\mathrm{bmax},l}^{(k)}\boldsymbol{P}^{(k),\mathrm{H}}\right) \leq p_{\mathrm{bmax},l}^{(k)} \forall k = 1, \ldots, K; l = 1, \ldots, L$$
$$\text{s.t. } \sum_{k=1}^{K} \mathrm{diag}\left(\boldsymbol{P}^{(k)}\boldsymbol{P}^{(k),\mathrm{H}}\right) \leq \boldsymbol{p}_{\mathrm{sum}}$$

with a diagonal weight matrix $\boldsymbol{W}^{(k)} = \mathrm{diag}\left(w_1^{(k)}, \ldots, w_L^{(k)}\right)$ that is used to implement the sum-rate optimization.

The Lagrange function $\phi(\boldsymbol{P}^{(k)}, \boldsymbol{\mu}_{\mathrm{sum}}, \boldsymbol{\mu}_{\mathrm{mask}}^{(k)}, \boldsymbol{\mu}_{\mathrm{bmax}}^{(k)})$ reads as

$$\phi(\boldsymbol{P}^{(k)}, \boldsymbol{\mu}_{\mathrm{sum}}, \boldsymbol{\mu}_{\mathrm{mask}}^{(k)}, \boldsymbol{\mu}_{\mathrm{bmax}}^{(k)}) = \sum_{k=1}^{K} \mathrm{tr}\left(\boldsymbol{W}^{(k)}\boldsymbol{e}^{(k)}\boldsymbol{e}^{(k),\mathrm{H}}\right) \tag{3.98}$$
$$+ \sum_{k=1}^{K}\sum_{l=1}^{L} \mu_{\mathrm{mask},l}^{(k)}\left(\mathrm{tr}\left(\boldsymbol{A}_l\boldsymbol{P}^{(k)}\boldsymbol{P}^{(k),\mathrm{H}}\right) - p_{\mathrm{mask},l}^{(k)}\right)$$
$$+ \sum_{k=1}^{K}\sum_{l=1}^{L} \mu_{\mathrm{bmax},l}^{(k)}\left(\mathrm{tr}\left(\boldsymbol{A}_l\boldsymbol{P}^{(k)}\boldsymbol{A}_{\mathrm{bmax},l}\boldsymbol{P}^{(k),\mathrm{H}}\right) - p_{\mathrm{bmax},l}^{(k)}\right)$$
$$+ \sum_{l=1}^{L} \mu_{\mathrm{sum},l}\left(\sum_{k=1}^{K} \mathrm{tr}\left(\boldsymbol{A}_l\boldsymbol{P}^{(k)}\boldsymbol{P}^{(k),\mathrm{H}}\right) - p_{\mathrm{sum},l}\right)$$

with the Lagrange multipliers $\boldsymbol{\mu}_{\mathrm{sum}}, \boldsymbol{\mu}_{\mathrm{bmax}}^{(k)}$ and $\boldsymbol{\mu}_{\mathrm{mask}}^{(k)}$ and the selection matrix $\boldsymbol{A}_l = \boldsymbol{e}_l\boldsymbol{e}_l^{\mathrm{T}}$ where $\boldsymbol{e}_l$ is the unit vector into dimension l. The matrix $\boldsymbol{A}_{\mathrm{bmax},l}^{(k)}$ is given by $\boldsymbol{A}_{\mathrm{bmax},l}^{(k)} = \boldsymbol{h}_l\boldsymbol{h}_l^H$ where $\boldsymbol{h}_l$ is the lth row of the channel matrix and the value $p_{\mathrm{bmax},l}$ is given by $p_{\mathrm{bmax},l} = \left(2^{b_{\max}} - 1\right)\Gamma\sigma^2$.

The first derivative of (3.98), the LMS step for the MMSE precoder, is

$$\frac{\partial\phi(\boldsymbol{P}^{(k)}, \boldsymbol{\mu}_{\mathrm{sum}}, \boldsymbol{\mu}_{\mathrm{mask}}^{(k)}, \boldsymbol{\mu}_{\mathrm{bmax}}^{(k)})}{\partial\boldsymbol{P}^{(k),*}} = \boldsymbol{H}^{(k),\mathrm{H}}\boldsymbol{G}^{(k),\mathrm{H}}\boldsymbol{W}^{(k)}\boldsymbol{e}^{(k)}\boldsymbol{u}^{(k),\mathrm{H}} \tag{3.99}$$
$$+ \mathrm{diag}\left(\boldsymbol{\mu}_{\mathrm{mask}}^{(k)} + \boldsymbol{\mu}_{\mathrm{sum}}\right)\boldsymbol{P}^{(k)} + \boldsymbol{P}^{(k)}\left(\sum_{l=1}^{L} \mu_{\mathrm{bmax},l}^{(k)}\boldsymbol{A}_{\mathrm{bmax},l}^{(k)}\right)$$

which gives the precoder update rule to be

$$\boldsymbol{P}^{(k),[t+1]} = \boldsymbol{P}^{(k),[t]} - \alpha_P\left(\frac{\partial\phi(\boldsymbol{P}^{(k)}, \boldsymbol{\mu}_{\mathrm{sum}}, \boldsymbol{\mu}_{\mathrm{mask}}^{(k)}, \boldsymbol{\mu}_{\mathrm{bmax}}^{(k)})}{\partial\boldsymbol{P}^{(k),*}}\right). \tag{3.100}$$

The Lagrange multipliers $\boldsymbol{\mu}_{\mathrm{mask}}^{(k)}$ which are used to control the transmit PSD, are obtained by another gradient step according to

$$\boldsymbol{\mu}_{\mathrm{mask}}^{(k),[t+1]} = \max\left(\boldsymbol{\mu}_{\mathrm{mask}}^{(k),[t]} + \alpha_{\mathrm{mask}}\left(\mathrm{diag}\left(\boldsymbol{P}^{(k)}\boldsymbol{P}^{(k),\mathrm{H}}\right) - \boldsymbol{p}_{\mathrm{mask}}^{(k)}\right), 0\right) \tag{3.101}$$

and the diagonal weight matrix $\boldsymbol{W}^{(k)} = \mathrm{diag}\left(w_1^{(k)}, \ldots, w_L^{(k)}\right)$ is defined to be the inverse of the mean squared error

$$w_l^{(k)} = \mathrm{E}\left[\Gamma|e_l^{(k)}|^2\right]^{-1} \tag{3.102}$$

and α_{mask} is the step size for the update of the Lagrangian.

The sum-power Lagrangian multipliers are updated according to

$$\boldsymbol{\mu}_{\mathrm{sum}}^{[t+1]} = \max\left(\boldsymbol{\mu}_{\mathrm{sum}}^{[t]} + \alpha_{\mathrm{sum}}\left(\mathrm{diag}\left(\sum_{k=1}^{K}\boldsymbol{P}^{(k)}\boldsymbol{P}^{(k),\mathrm{H}}\right) - \boldsymbol{p}_{\mathrm{sum}}\right), 0\right) \tag{3.103}$$

with the step size α_{sum}. Finally, the bit allocation upper bound multipliers $\boldsymbol{\mu}_{\mathrm{bmax}}^{(k)}$ are updated according to

$$\mu_{\mathrm{bmax},l}^{(k),[t+1]} = \max\left(\mu_{\mathrm{bmax},l}^{(k),[t]} + \alpha_{\mathrm{bmax}}\left(SNR_l^{(k)} - (2^{b_{\max}} - 1)\Gamma\right), 0\right) \tag{3.104}$$

with the step size α_{bmax}.

The precoder update rule consisting of Eqs. (3.100)–(3.104) does not require matrix inversion, which is an advantage for implementation. While for ZF precoding, two adaptive optimization methods run in parallel, one for the precoder and one for the transmit spectrum, this is not case for MMSE precoding. The MMSE precoder update implicitly satisfies the power constraints.

While linear precoding methods in DSL/G.fast systems are well understood, high crosstalk of G.fast channels results in an increased gap to channel capacity due to linear precoding. This is a potential application of nonlinear precoding methods.

3.3 Downstream Nonlinear Precoding

The first proposals on VDSL2 crosstalk cancelation in [44, 45] brought up the idea to use Tomlinson Harashima precoding (THP) in the downstream direction. The original THP papers [46, 47] discuss time domain channel equalization with nonlinear precoding. In [48], it is shown that the results can be transferred from time domain intersymbol interference cancelation to MIMO crosstalk cancelation.

For VDSL2 crosstalk cancelation, the idea of THP did not come into practice because of increased complexity that gives only small performance advantages over

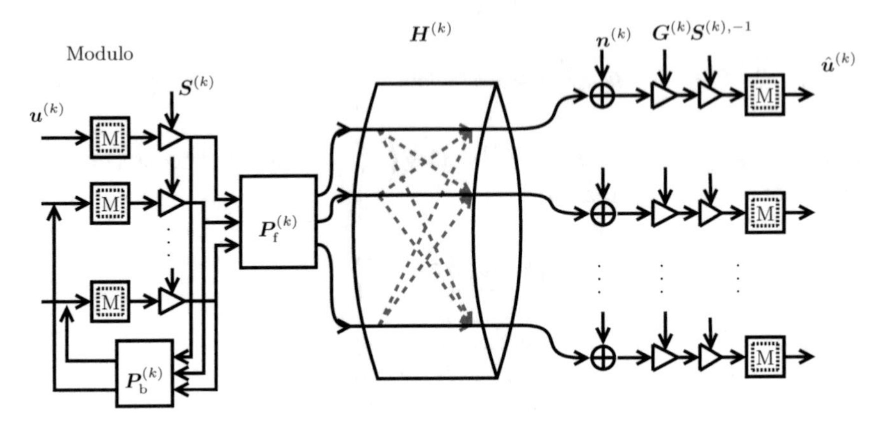

Fig. 3.14 Tomlinson–Harashima precoding model representing one subcarrier

linear zero-forcing methods [21]. While the arguments in [21] hold true for VDSL2 channels, the crosstalk conditions change for the high frequency spectrum of G.fast up to 106 or 212 MHz. Therefore, Tomlinson–Harashima precoding is analyzed as an alternative precoding approach for G.fast.

The system model for one subcarrier and THP in downstream direction is shown in Fig. 3.14. The transmitter as well as the receivers contain a modulo operation block as described in Eq. (3.9).

The precoder consists of the modulo operations mod (.) and a feedback part, using the strictly lower triangular feedback matrix $\boldsymbol{P}_\mathrm{b}^{(k)} \in \mathbb{C}^{L\times L}$ and the spectrum shaping matrix $\boldsymbol{S}^{(k)}$. The modulo output signals are processed with another matrix $\boldsymbol{P}_\mathrm{f}^{(k)} \in \mathbb{C}^{L\times L}$.

At the receiver side, another modulo operation is required after the diagonal equalizer matrix $\boldsymbol{G}^{(k)} \in \mathbb{C}^{L\times L}$ and the inverse gain scaling. Reference [45] describes the nonlinear precoder structure in more detail.

The precoder output signal $\boldsymbol{y}^{(k)}$ is determined by the following steps

$$\boldsymbol{u}_\mathrm{back}^{(k)} = \boldsymbol{S}^{(k),-1}\boldsymbol{P}_\mathrm{b}^{(k)}\boldsymbol{u}_\mathrm{mod}^{(k)}, \tag{3.105}$$

$$\boldsymbol{u}_\mathrm{mod}^{(k)} = \boldsymbol{S}^{(k)} \mod \left(\boldsymbol{u}_\mathrm{back}^{(k)} + \boldsymbol{u}^{(k)}\right), \tag{3.106}$$

$$\boldsymbol{y}^{(k)} = \boldsymbol{P}_\mathrm{f}^{(k)}\boldsymbol{u}_\mathrm{mod}^{(k)} \tag{3.107}$$

where $\boldsymbol{u}_\mathrm{back}^{(k)}$ is the feedback signal after multiplication with the feedback matrix $\boldsymbol{P}_\mathrm{b}^{(k)}$ and $\boldsymbol{u}_\mathrm{mod}^{(k)}$ is the output signal of the modulo operations according to Eq. (3.9).

The receive signal vector $\hat{\boldsymbol{u}}^{(k)}$ is given by

$$\hat{\boldsymbol{u}}^{(k)} = \mod \left(\boldsymbol{S}^{(k),-1}\boldsymbol{G}^{(k)}\left(\boldsymbol{H}^{(k)}\boldsymbol{y}^{(k)} + \boldsymbol{n}^{(k)}\right)\right). \tag{3.108}$$

The feedback loop in the nonlinear precoder implies an encoding order, which has some influence on the performance of each line. The line, which is encoded first, experiences the smallest performance penalty due to precoding, which increases towards the line which is encoded last. The encoding order may be different for each carrier k.

3.3.1 Precoding Losses of THP

Tomlinson Harashima precoding is an approximation of the optimal dirty paper coding scheme [49]. It is not capacity achieving due to different precoding losses, but achieves higher performance than linear precoding under certain conditions such as high crosstalk. An understanding of THP losses is very important to justify the cases where THP is beneficial [7]. There are three types of THP losses, the shaping loss, the power loss and the modulo loss [14]. Besides that, the impact of channel estimation accuracy and quantization losses are discussed in [50].

Shaping Loss

Channel capacity is achieved by the use' of Gaussian transmit signals. At the output of the THP, transmit signals have a uniform distribution within the modulo region, which is suboptimal. The difference between the channel capacity and the achievable rate based on uniformly distributed signals is called the shaping loss. The shaping loss depends on the signal-to-noise ratio and increases with increasing SNR. It reaches the ultimate limit of 1.53 dB in the high-SNR region. Several approaches have been made to reduce the shaping loss in THP, e.g., [51].

Implementable transmission systems like G.fast do not use Gaussian signals, but use QAM modulated signals as discussed in Sect. 3.1.4. Therefore, the shaping loss is present for both, linear and nonlinear precoding in G.fast.

Power Loss

The variance of the modulo output signal $\boldsymbol{u}_{\mathrm{mod}}^{(k)}$ is higher than the variance of the desired signal $\boldsymbol{u}^{(k)}$. The increased power is the power loss [14]. The power loss depends on the used constellation size and it achieves the maximum of 1.25 dB for 1-bit and 2-bit constellations. Besides the constellation size, the actual power loss also depends on the crosstalk conditions and the encoding order. The line which is encoded first does not experience any power loss. With increasing number of encoded lines, the power loss increases [52].

The power loss cannot be ignored in G.fast systems as it may cause a violation of the transmit power constraints. Assuming the covariance matrix of the modulo output signal to be $\boldsymbol{C}_{u_{\mathrm{mod}}u_{\mathrm{mod}}} = \mathrm{E}\left[\boldsymbol{u}_{\mathrm{mod}}^{(k)}\boldsymbol{u}_{\mathrm{mod}}^{(k),\mathrm{H}}\right]$, a simplified model of the power loss assumes that the variance of the signal $\boldsymbol{u}_{\mathrm{mod}}$ is given by

$$\left[\boldsymbol{C}_{u_{\mathrm{mod}}u_{\mathrm{mod}}}^{(k)}\right]_{ll} = p_{\mathrm{m},l}^{(k),2} x_l^{(k)} \tag{3.109}$$

Table 3.1 Worst case power loss per G.fast constellation

Constellation $\hat{b}$	1	2	3	4	5	6
Power loss $p_{\mathrm{m},l}^{(k)}$/dB	1.25	1.25	0.28	0.28	0.80	0.06
Constellation $\hat{b}$	7	8	9	10	11	12
Power loss $p_{\mathrm{m},l}^{(k)}$/dB	0.68	0.02	0.66	0	0.65	0

with the power loss described by a scalar gain $p_{\mathrm{m},l}^{(k)}$ for line l and carrier k. The power loss for G.fast constellations is analyzed in [15] which shows the worst case power loss according to Table 3.1.

The power loss depends on the encoding order. For the first line encoded, it is 0 dB and increases with each encoded line towards the worst case values in Table 3.1. The actual power loss can be derived from the the measured covariance matrix $\boldsymbol{C}_{u_{\mathrm{mod}}u_{\mathrm{mod}}}^{(k)}$ using Eq. (3.109), which gives less conservative values than Table 3.1 and allows some performance improvement.

Modulo Loss

This third loss component of THP, the modulo loss, also takes its maximum for the 1- and 2-bit constellations. The modulo loss is a result of the modulo operation that is required at the receiver (see Fig. 3.14).

The modulo loss is discussed in Sect. 3.1.4. Appendix B.2 summarizes the modulo loss obtained from simulations of uncoded QAM modulation. Appendices B.3 and B.4 show the corresponding results for trellis coded modulation and trellis and Reed–Solomon coding.

3.3.2 *Nonlinear Zero-Forcing Precoding*

For nonlinear zero-forcing, the precoder matrices $\boldsymbol{P}_{\mathrm{f}}^{(k)}$ and $\boldsymbol{P}_{\mathrm{b}}^{(k)}$ are derived by QR decomposition [45].

Starting with the QR decomposition of the transpose channel matrix

$$\boldsymbol{Q}^{(k)}\boldsymbol{R}^{(k)} = \boldsymbol{H}^{(k),\mathrm{T}} \tag{3.110}$$

into the unitary matrix $\boldsymbol{Q}^{(k)}$ and the upper triangular matrix $\boldsymbol{R}^{(k)}$, the precoder matrices are derived as follows. The forward matrix is given by

$$\boldsymbol{P}_{\mathrm{f}}^{(k)} = \boldsymbol{Q}^{(k),*} \tag{3.111}$$

and the lower triangular feedback matrix $\boldsymbol{P}_{\mathrm{b}}^{(k)}$ is

$$\boldsymbol{P}_{\mathrm{b}}^{(k)} = \left(\boldsymbol{I}_L - \mathrm{diag}\left(\mathrm{diag}\left(\boldsymbol{R}^{(k)}\right)\right)^{-1} \boldsymbol{R}^{(k),\mathrm{T}}\right). \tag{3.112}$$

The equalizer at receive side, which satisfies the zero-forcing condition, is obtained by

$$\boldsymbol{G}^{(k)} = \operatorname{diag}\left(\operatorname{diag}\left(\boldsymbol{R}^{(k)}\right)\right)^{-1}. \tag{3.113}$$

This definition of the precoder matrices encodes the lines in natural order from 1 to L. To create a different encoding order, the permutation matrix $\boldsymbol{\Pi}^{(k)}$ is introduced.

Assuming an encoding order vector $\boldsymbol{o}^{(k)}$ with the indices of the encoded lines to be $\boldsymbol{o}^{(k)} \in \{1, \ldots, L\}^L$, the permutation matrix is given by

$$\pi_{vd} = \begin{cases} 1 & \text{for } o_d = v \\ 0 & \text{otherwise} \end{cases}. \tag{3.114}$$

The precoder coefficients for THP with encoding order $\boldsymbol{o}^{(k)}$ is based on the QR decomposition according to

$$\boldsymbol{Q}^{(k)}\boldsymbol{R}^{(k)} = \boldsymbol{\Pi}\boldsymbol{H}^{(k),\mathrm{T}}\boldsymbol{\Pi}^{\mathrm{T}}. \tag{3.115}$$

The forward matrix is now given by

$$\boldsymbol{P}_{\mathrm{f}}^{(k)} = \boldsymbol{\Pi}^{\mathrm{T}}\boldsymbol{Q}^{(k),*}\boldsymbol{\Pi} \tag{3.116}$$

and the lower triangular feedback matrix is given by

$$\boldsymbol{P}_{\mathrm{b}}^{(k)} = \boldsymbol{\Pi}^{\mathrm{T}}\left(\boldsymbol{I}_L - \operatorname{diag}\left(\operatorname{diag}\left(\boldsymbol{R}^{(k)}\right)\right)^{-1}\boldsymbol{R}^{(k),\mathrm{T}}\right)\boldsymbol{\Pi}^{\mathrm{T}}. \tag{3.117}$$

The equalizer at receive side, which satisfies the zero-forcing condition, is obtained by

$$\boldsymbol{G}^{(k)} = \boldsymbol{\Pi}^{\mathrm{T}}\operatorname{diag}\left(\operatorname{diag}\left(\boldsymbol{R}^{(k)}\right)\right)^{-1}\boldsymbol{\Pi}. \tag{3.118}$$

For zero-forcing THP, two parameters are subject to optimization, the transmit spectrum, determined by the gain scaling $\boldsymbol{S}^{(k)}$ and the encoding order $\boldsymbol{o}^{(k)}$. Spectrum optimization for THP is mentioned in [45], but the proposed method is based on assumptions for the channel matrices which do not hold for G.fast channels. Reference [15] discusses a greedy approach to bit loading and spectrum optimization for G.fast.

3.3.3 *Spectrum Optimization for Zero-Forcing THP*

In Sects. 3.2.8 and 3.2.10, it is shown that convex optimization methods give an efficient solution for the spectrum optimization problem for G.fast. These results are applied to nonlinear precoding and this section discusses the required changes.

The power constraints to be considered are the per-line spectral mask and per-line sum-power constraint, as for linear precoding. The bit loading upper bound gives a third power constraint. The modulo loss $p_{\mathrm{m},l}^{(k)}$, the increase of transmit signal power due to the nonlinear precoding is incorporated into the power constraints by the diagonal modulo loss matrix $\boldsymbol{P}_{\mathrm{m}}^{(k)} = \operatorname{diag}\left(p_{\mathrm{m},1}^{(k)}, \ldots, p_{\mathrm{m},L}^{(k)}\right)$. The spectrum optimization constraint set for linear ZF precoding is given by

$$\bar{\boldsymbol{A}}^{(k)} = \begin{bmatrix} \left(\boldsymbol{P}_{\mathrm{f}}^{(k)}\boldsymbol{P}_{\mathrm{m}}^{(k)}\right) \odot \left(\boldsymbol{P}_{\mathrm{f}}^{(k)}\boldsymbol{P}_{\mathrm{m}}^{(k)}\right)^{*} \\ \boldsymbol{I}_L \\ -\boldsymbol{I}_L \end{bmatrix} \quad \boldsymbol{d}^{(k)} = \begin{bmatrix} \boldsymbol{p}_{\mathrm{mask}}^{(k)} \\ \boldsymbol{p}_{\mathrm{bmax}}^{(k)} \\ \boldsymbol{0}_L \end{bmatrix} \tag{3.119}$$

with the constraint matrix $\bar{\boldsymbol{A}}^{(k)}$ and the constraint vector $\boldsymbol{d}^{(k)}$. For the per-line sum-power constraint, the constraint matrix $\bar{\boldsymbol{A}}_{\mathrm{sum}}^{(k)}$ is modified according to

$$\bar{\boldsymbol{A}}_{\mathrm{sum}}^{(k)} = \left[\left(\boldsymbol{P}\mathrm{f}^{(k)}\boldsymbol{P}_{\mathrm{m}}^{(k)}\right) \odot \left(\boldsymbol{P}_{\mathrm{f}}^{(k)}\boldsymbol{P}_{\mathrm{m}}^{(k)}\right)^{*}\right] \tag{3.120}$$

which gives the gradient update for the sum-power Lagrangian multipliers to be

$$\boldsymbol{\mu}_{\mathrm{sum\ grad}}^{[t]} = \boldsymbol{\mu}_{\mathrm{sum}}^{[t]} + \alpha_{\mathrm{sum}}\left(\sum_{k=1}^{K} \bar{\boldsymbol{A}}_{\mathrm{sum}}^{(k)}\boldsymbol{x}^{(k)} - \boldsymbol{p}_{\mathrm{sum}}\right). \tag{3.121}$$

With the constraint set of Eqs. (3.119) and (3.120), Algorithm 9 as well as Algorithm 5 can be applied to linear and non-linear precoding. The modulo loss is considered in the rate function by a change of the SNR gap Γ, while the idea of Sect. 3.2.8 to approximate the discrete rate function by a continuous rate function still holds. The sum-rate optimization problem for ZF THP reads as

$$\begin{aligned} \max_{\boldsymbol{x}^{(1)},\ldots,\boldsymbol{x}^{(K)}} \sum_{l=1}^{L} R_l \text{ s.t. } \bar{\boldsymbol{A}}^{(k)}\boldsymbol{x}^{(k)} \leq \boldsymbol{d}^{(k)}\ \forall\, k = 1, \ldots, K \\ \text{s.t. } \sum_{k=1}^{K} \bar{\boldsymbol{A}}_{\mathrm{sum}}^{(k)}\boldsymbol{x}^{(k)} \leq \boldsymbol{p}_{\mathrm{sum}}, \end{aligned} \tag{3.122}$$

with the rate function according to Eq. (3.60) [45].

As $\boldsymbol{P}_{\mathrm{f}}$ is a unitary matrix, which is always well defined and $\boldsymbol{P}_{\mathrm{m}}^{(k)}$ is close to the identity matrix, the constraint set for Tomlinson–Harashima precoding, Eq. (3.119), is less restrictive than the constraint set for linear precoding according to Eq. (3.56), which gives a performance advantage for nonlinear precoding. This comes for the price of the modulo loss and the increased precoding complexity.

Algorithm 6 summarizes the spectrum optimization algorithm for nonlinear ZF precoding.

Algorithm 6 Spectrum optimization for zero-forcing THP

Initialize $\boldsymbol{\mu}_{\text{sum}} = \mathbf{0}_L$
repeat
  for $k = 1$ **to** K **do**
    Start with all lines active $\mathbb{I}_{\text{a}}^{(k)} = \{1, \ldots, L\}$
    Initialize power loss matrix $\boldsymbol{P}_{\text{m}}^{(k)}$ with worst-case approximation according to Table 3.1
    repeat
      Calculate ZF precoder $\boldsymbol{P}_{\text{b}}^{(k)}$ and $\boldsymbol{P}_{\text{f}}^{(k)}$ for given $\mathbb{I}_{\text{a}}$ using Eqs. (3.117) and (3.116)
      Initialize $\boldsymbol{x}^{(k)}$
      Calculate constraint set Eqs. (3.55) and (3.119)
      repeat
        Calculate quadratic approximation using (3.63) and (3.64) and solve (3.66)
      until Convergence of $\boldsymbol{x}^{(k)}$
      if $\exists SNR_l^{(k)} < SNR_{\min} : l \in \mathbb{I}_{\text{a}}^{(k)}$ **then**
        Remove the weakest link $\arg\max_{l \in \mathbb{I}_{\text{a}}} \left|g_l^{(k)}\right|^2$ from the set of active lines
      end if
    until All active lines/carriers meet SNR lower bound
  end for
  Update power allocation using (3.121)
until Per-line sum-power is converged

Algorithm 6 is suited for precoder initialization. For tracking, the algorithm from Sect. 3.2.10 can be used. The corresponding algorithm is described in Algorithm 12 in Appendix B.11.

Similar to linear weighted MMSE precoding, it is possible to allow interference for THP. The corresponding methods to derive the MMSE TH precoder and equalizer matrices are described in [53]. Due to the algorithmic complexity, they are not suited for a G.fast implementation, but MMSE THP gives an upper bound to the achievable rates of nonlinear precoding in general. These upper bounds are investigated in Sect. 3.5.

3.4 Upstream Optimization

In upstream direction, receiver coordination at the DPU side is used to suppress crosstalk. The CPE transmitters send the uplink signals independently. Determining the receive equalizer matrices is an unconstrained optimization problem. The transmit spectrum optimization can be solved as described in Sect. 3.1.6 for a given equalizer matrix $\boldsymbol{G}^{(k)}$. The upstream system model is shown in Fig. 3.15 for one subcarrier.

The receive signal $\hat{\boldsymbol{u}}^{(k)}$ is given by

$$\hat{\boldsymbol{u}}^{(k)} = \boldsymbol{G}^{(k)} \left(\boldsymbol{H}^{(k)}\boldsymbol{S}^{(k)}\boldsymbol{u}^{(k)} + \boldsymbol{n}^{(k)}\right). \tag{3.123}$$

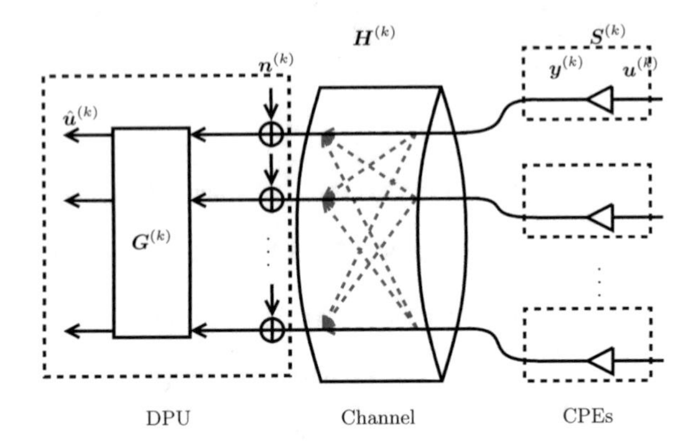

Fig. 3.15 System model for one carrier in upstream direction

In upstream direction, only linear methods are considered for G.fast. There are two main reasons. In downstream direction, linear precoding may cause an increase of the transmit power and suboptimal spectrum shaping methods cause significant performance losses compared to the crosstalk-free data rates. This is less severe in upstream direction. Furthermore, the main target application for G.fast are consumer services with priority on downlink rate. For these applications, upstream performance is less critical.

3.4.1 Zero Forcing Equalization

Again, there are two approaches to determine the precoder coefficients. Zero-forcing equalization is based on the inversion of the channel matrix

$$G^{(k)} = \left(H^{(k)} S^{(k)}\right)^{-1} \tag{3.124}$$

where the inverse gain scaling at the receiver $S^{(k),-1}$ is incorporated into the equalizer matrix $G^{(k)}$.

Similar to zero-forcing precoding in downlink, there are cases where some of the active lines do not use a certain carrier because the SNR is not sufficient to transmit at least one bit. Assuming that $\mathbb{I}_{\mathrm{a}}^{(k)} \subset 1, \ldots, L$ lines use carrier k, the equalizer has dimensions $G^{(k)} \in \mathbb{C}^{L_{\mathrm{a}}^{(k)} \times L}$ where $L_{\mathrm{a}}^{(k)} = \left|\mathbb{I}_{\mathrm{a}}^{(k)}\right|$. In this case, the equalizer matrix is the pseudoinverse of the reduced channel $H^{(k)} \in \mathbb{C}^{L \times L_{\mathrm{a}}^{(k)}}$ according to

$$G^{(k)} = \left[H^{(k)} S^{(k)}\right]^{+} \tag{3.125}$$

where $[]^{+}$ denotes the Moore–Penrose pseudoinverse.

Assuming that the zero-forcing condition is satisfied, the receive error $\boldsymbol{e}^{(k)}$ is given by

$$\boldsymbol{e}^{(k)} = \hat{\boldsymbol{u}}^{(k)} - \boldsymbol{u}^{(k)} = \boldsymbol{G}^{(k)}\boldsymbol{n}^{(k)}. \tag{3.126}$$

The receive error covariance matrix $\boldsymbol{C}_{ee}^{(k)} = \mathrm{E}\left[\boldsymbol{e}^{(k)}\boldsymbol{e}^{(k),\mathrm{H}}\right]$ is given by

$$\boldsymbol{C}_{ee}^{(k)} = \boldsymbol{G}^{(k)}\boldsymbol{G}^{(k),\mathrm{H}}\sigma^2 \tag{3.127}$$

which gives the signal-to-noise ratio $SNR_l^{(k)}$ on carrier k and line l to be

$$SNR_l^{(k)} = \frac{1}{\left[\boldsymbol{C}_{ee}^{(k)}\right]_{ll}}. \tag{3.128}$$

Algorithm 7 summarizes the initialization of the ZF equalizer.

Algorithm 7 Linear Zero-Forcing Equalizer

Initialize
for $k = 1$ **to** K **do**
  Start with all lines active $L_{\mathrm{a}}^{(k)} = 1, \ldots, L$
  repeat
    Calculate ZF equalizer (3.125)
    Calculate SNR for each line
    if There is at least one $SNR_i^{(k)} < SNR_{\min}$ **then**
      Remove the line with the lowest SNR from $\mathbb{I}_{\mathrm{a}}^{(k)}$
      Update selected active channels $\mathbb{I}_{\mathrm{a}}^{(k)}$
    end if
  until All active carriers meet SNR lower bound
end for

The main issue of ZF equalization is the fact that with an ill-conditioned channel matrix $\boldsymbol{H}^{(k)}$, the precoder matrix $\boldsymbol{G}^{(k)}$ may result in a significant noise increase. The issue is partially resolved by removing such subcarriers from the active lines set $\mathbb{I}_{\mathrm{a}}^{(k)}$. While this is based on a hard decision to enable or disable certain carriers, MMSE equalization gives a more generalized approach to minimize the receiver error.

3.4.2 *MMSE Equalization*

The idea of the MMSE equalizer in upstream, similar to the MMSE precoder, is to allow interference in case that it improves the overall performance. Upstream MMSE equalization can be implemented with lower effort and lower complexity than MMSE precoding. Therefore, it is common in G.fast or VDSL2 upstream crosstalk canceler implementations.

The general receiver error $\boldsymbol{e}^{(k)}$ is given by

$$\boldsymbol{e}^{(k)} = \hat{\boldsymbol{u}}^{(k)} - \boldsymbol{u}^{(k)} = \boldsymbol{G}^{(k)} \left(\boldsymbol{H}^{(k)} \boldsymbol{S}^{(k)} \boldsymbol{u}^{(k)} + \boldsymbol{n}^{(k)} \right) - \boldsymbol{u}^{(k)}. \tag{3.129}$$

which gives the MMSE equalizer $\boldsymbol{G}^{(k)}$ to be

$$\boldsymbol{G}^{(k)} = \boldsymbol{H}^{(k),\mathrm{H}} \left(\boldsymbol{H}^{(k)} \boldsymbol{S}^{(k),2} \boldsymbol{H}^{(k),\mathrm{H}} + \sigma^2 \boldsymbol{I} \right)^{-1}. \tag{3.130}$$

The optimal equalizer depends on the transmit scaling $\boldsymbol{S}^{(k)}$ which results in an iterative optimization process of consecutive spectrum and precoder optimization steps. While Eq. (3.130) is used to initialize the equalizer, the equalizer is updated in the tracking phase with an LMS update step.

The descending direction of error is given by

$$\frac{\partial \mathrm{tr} \left(\boldsymbol{e}^{(k)} \boldsymbol{e}^{(k),\mathrm{H}} \right)}{\partial \boldsymbol{G}^{(k),*}} = \boldsymbol{e}^{(k)} \hat{\boldsymbol{y}}^{(k),\mathrm{H}}$$

where $\hat{\boldsymbol{y}}^{(k)} = \boldsymbol{H}^{(k)} \boldsymbol{S}^{(k)} \boldsymbol{u}^{(k)} + \boldsymbol{n}^{(k)}$ is the signal vector at the DPU receiver.

The corresponding LMS update step with step size α_G is

$$\boldsymbol{G}^{(k),[t+1]} = \boldsymbol{G}^{(k),[t]} - \alpha_G \boldsymbol{e}^{(k),[t]} \hat{\boldsymbol{y}}^{(k)[t],\mathrm{H}}. \tag{3.131}$$

The described methods can be used for initialization and tracking for the upstream equalizer. For a line joining, there is a low complexity approach towards the zero-forcing equalizer coefficients, which is described in Appendix B.8.

3.5 Theoretical Limits

While the precoding and spectrum optimization methods discussed in Sects. 3.2 and 3.3 focus on low complexity implementable methods, this section gives more insight in the capacity limits of the G.fast channel. The results show the remaining gap between the achieved rates of implementable methods and the upper bounds.

The capacity analysis extends the work in [1] and shows the impact of implementation losses due to the bit loading upper bound and the imperfect coding in G.fast. In addition to [1], the capacity analysis is done with both, the per-line sum-power constraint and the per-line spectral mask constraint and includes different cable types.

Most results on capacity of the MIMO broadcast channel consider only a sum-power constraint. In [54] the capacity of the MIMO broadcast channel with a more realistic per-antenna constraint is derived, using uplink-downlink duality. An interior point method is proposed in [54] to calculate the capacity region. For G.fast, where a per-line spectral mask constraint and a per-line sum-power constraint is considered, the algorithm is not suitable.

More recent results in [55] present a method to compute channel capacity of the G.fast channel. The iterative approach in [55] is used as a basis for this investigation.

Capacity is achieved by the dirty paper coding (DPC) scheme [49]. The capacity is given by maximizing

$$C = \sum_{k=1}^{K} \sum_{l=1}^{L} \log_2 \left(1 + \frac{x_l^{(k)} \left| \boldsymbol{h}_l^{(k),\mathrm{T}} \boldsymbol{p}_l^{(k)} \right|^2}{\Gamma \left(\sigma^2 + \sum_{d>l} x_d^{(k)} \left| \boldsymbol{h}_l^{(k),\mathrm{T}} \boldsymbol{p}_d^{(k)} \right|^2 \right)} \right), \tag{3.132}$$

where the channel matrix $\boldsymbol{H}^{(k)}$ is partitioned into row vectors $\boldsymbol{h}_l^{(k),\mathrm{T}}$ for each line l and the precoder matrix is partitioned into column vectors $\boldsymbol{p}_l^{(k)}$. The noise variance is σ^2 and Γ is the SNR gap.

Reference [55] formulates the capacity maximization problem with two power constraints, the per-line sum-power constraint and the per-line spectral mask constraint. The results are extended to include the bit loading upper bound and an SNR gap $\Gamma > 1$, which gives

$$\begin{aligned}
\max_{\boldsymbol{P}^{(k)}, \boldsymbol{x}^{(k)}} C \quad & \text{s.t. diag} \left(\sum_{k=1}^{K} \boldsymbol{P}^{(k)} \mathrm{diag}(\boldsymbol{x}^{(k)}) \boldsymbol{P}^{(k),\mathrm{H}} \right) \le \boldsymbol{p}_{\mathrm{sum}} \\
& \text{s.t. diag} \left(\boldsymbol{P}^{(k)} \mathrm{diag}(\boldsymbol{x}^{(k)}) \boldsymbol{P}^{(k),\mathrm{H}} \right) \le \boldsymbol{p}_{\mathrm{mask}} \; \forall\, k = 1, \ldots, K \\
& \text{s.t.} \frac{x_l^{(k)} \left| \boldsymbol{h}_l^{(k),\mathrm{T}} \boldsymbol{p}_l^{(k)} \right|^2}{\left(\sigma^2 + \sum_{d>l} x_d^{(k)} \left| \boldsymbol{h}_l^{(k),\mathrm{T}} \boldsymbol{p}_d^{(k)} \right|^2 \right)} \le \Gamma (2^{b_{\max}} - 1) \; \forall \begin{array}{l} k = 1, \ldots, K, \\ l = 1, \ldots, L \end{array}
\end{aligned} \tag{3.133}$$

The Lagrangian function $\Phi(\boldsymbol{x}^{(k)}, \boldsymbol{\mu}_{\mathrm{sum}}, \boldsymbol{\mu}_{\mathrm{mask}}^{(k)}, \boldsymbol{\mu}_{\mathrm{bmax}}^{(k)})$ corresponding to Eq. (3.133) reads as

$$\begin{aligned}
& \Phi(\boldsymbol{x}^{(k)}, \boldsymbol{\mu}_{\mathrm{sum}}, \boldsymbol{\mu}_{\mathrm{mask}}^{(k)}, \boldsymbol{\mu}_{\mathrm{bmax}}^{(k)}) = -C \\
& + \boldsymbol{\mu}_{\mathrm{sum}}^{\mathrm{T}} \left(\mathrm{diag} \left(\sum_{k=1}^{K} \boldsymbol{P}^{(k)} \mathrm{diag}(\boldsymbol{x}^{(k)}) \boldsymbol{P}^{(k),\mathrm{H}} \right) - \boldsymbol{p}_{\mathrm{sum}} \right) \\
& + \sum_{k=1}^{K} \boldsymbol{\mu}_{\mathrm{mask}}^{(k),\mathrm{T}} \left(\mathrm{diag} \left(\boldsymbol{P}^{(k)} \mathrm{diag}(\boldsymbol{x}^{(k)}) \boldsymbol{P}^{(k),\mathrm{H}} \right) - \boldsymbol{p}_{\mathrm{mask}} \right) \\
& + \sum_{k=1}^{K} \sum_{l=1}^{L} \mu_{\mathrm{bmax},l}^{(k)} \left(x_l^{(k)} \frac{\left| \boldsymbol{h}_l^{(k),\mathrm{T}} \boldsymbol{p}_l^{(k)} \right|^2}{\sigma^2} - \Gamma (2^{b_{\max}} - 1) \right)
\end{aligned} \tag{3.134}$$

where $\boldsymbol{\mu}_{\text{sum}}$, $\boldsymbol{\mu}_{\text{mask}}^{(k)}$ and $\boldsymbol{\mu}_{\text{bmax}}^{(k)}$ are the Lagrange multipliers of the per-line sum-power constraint, the spectral mask constraint and the bit loading upper bound, respectively.

In Eq. (3.134), the zero-forcing approximation of the SNR upper bound is used, assuming $\sum_{d>l} x_d |\boldsymbol{h}_l^{(k),\text{T}} \boldsymbol{p}_d^{(k)}|^2 \ll \sigma^2$ for $SNR_l^{(k)} \geq \Gamma(2^{b_{\max}} - 1)$. The necessary conditions for the approximation are explained in Appendix B.12. According to the explanation in Sect. 3.2.9, this approximation allows for a convex optimization problem in the dual uplink.

Following [55], the solution of Eq. (3.133) is obtained using an uplink-downlink duality approach. The dual uplink channel is $\boldsymbol{H}_{\text{u}}^{(k)} = \boldsymbol{H}^{(k),\text{H}}$ and the receiver noise covariance matrix is $\boldsymbol{Q}^{(k)}$, which is equivalent to the uplink-downlink duality approach used for linear MMSE precoding as described in Sect. 3.2.9 and Fig. 3.13. Furthermore, the uplink power allocation for line l and carrier k is $x_{\text{u},l}^{(k)}$, which corresponds to the dual uplink precoder in Fig. 3.13, e.g., $x_{\text{u},l}^{(k)} = |p_{\text{u},ll}|^2$.

Hereby, the dual uplink equalizers $\boldsymbol{G}_{\text{u}}^{(k)} = \boldsymbol{P}^{(k),\text{H}}$ are given by

$$\boldsymbol{p}_l^{(k),\text{H}} = \boldsymbol{g}_{\text{u},l}^{(k),\text{T}} = \beta_l^{(k)} \sqrt{x_{\text{u},l}^{(k)}} \boldsymbol{h}_{\text{u},l}^{(k),\text{H}} \Big(\text{diag}(\boldsymbol{\mu}_{\text{sum}} + \boldsymbol{\mu}_{\text{mask}}^{(k)}) + \boldsymbol{H}_{\text{u}}^{(k)} \text{diag}(\boldsymbol{\mu}_{\text{bmax}}^{(k)}) \boldsymbol{H}_{\text{u}}^{(k),\text{H}} + \sum_{d \leq l} x_{\text{u},d}^{(k)} \boldsymbol{h}_{\text{u},d}^{(k)} \boldsymbol{h}_{\text{u},d}^{(k),\text{H}} \Big)^{-1} \tag{3.135}$$

where the scaling factor $\beta_l^{(k)}$ is chosen to satisfy $||\boldsymbol{g}_{\text{u},l}^{(k)}||_2^2 = 1$. $\boldsymbol{g}_{\text{u},l}^{(k),\text{T}}$ is the lth row vector of the dual uplink equalizer matrix.

The dual uplink power power $\boldsymbol{x}_{\text{u}}^{(k)}$ is obtained by a fixed point iteration according to

$$x_{\text{u},l}^{(k)} = \frac{\log 2}{1 + \sum\limits_{d>l} \frac{\log 2}{\Gamma + x_{\text{u},d}^{(k)} \boldsymbol{h}_{\text{u},d}^{(k),\text{H}} \boldsymbol{A}_d^{(k),-1} \boldsymbol{h}_{\text{u},d}^{(k)}} |\boldsymbol{h}_{\text{u},l}^{(k),\text{H}} \boldsymbol{A}_d^{(k),-1} \boldsymbol{h}_{\text{u},l}^{(k)}|^2} \tag{3.136}$$

where $\boldsymbol{A}_d^{(k)}$ is given by

$$\boldsymbol{A}_d^{(k)} = \text{diag}(\boldsymbol{\mu}_{\text{sum}} + \boldsymbol{\mu}_{\text{mask}}^{(k)}) + \boldsymbol{H}_{\text{u}}^{(k)} \text{diag}(\boldsymbol{\mu}_{\text{bmax}}^{(k)}) \boldsymbol{H}_{\text{u}}^{(k),\text{H}} + \sum_{d<l} x_{\text{u},d}^{(k)} \boldsymbol{h}_{\text{u},d}^{(k)} \boldsymbol{h}_{\text{u},d}^{(k),\text{H}}. \tag{3.137}$$

To guarantee positiveness of the dual uplink power, the projection

$$x_{\text{u},l}^{(k)} = \max(x_{\text{u},l}^{(k)}, 0) \tag{3.138}$$

is performed in each step of the fixed point iteration.

The downlink precoder $\boldsymbol{P}^{(k)}$ is obtained from $\boldsymbol{G}_{\text{u}}^{(k)}$ according to

$$\boldsymbol{P}^{(k)} = \boldsymbol{G}_{\text{u}}^{(k),\text{H}}. \tag{3.139}$$

The transformation from the dual uplink power allocation to the downlink power allocation is performed according to [55] by the matrix $\boldsymbol{Z}^{(k)}$,

$$\boldsymbol{Z}^{(k)}\boldsymbol{x}^{(k)} = \boldsymbol{x}_{\mathrm{u}}^{(k)} \tag{3.140}$$

where $\boldsymbol{Z}^{(k)}$ is given by

$$z_{vd} = \begin{cases} \boldsymbol{p}_v^{(k),\mathrm{H}}\boldsymbol{Q}^{(k)}\boldsymbol{p}_v^{(k)} + \sum_{l<v} x_{\mathrm{u},l}^{(k)} \left|\boldsymbol{p}_d^{(k),\mathrm{H}}\boldsymbol{h}_{\mathrm{u},l}^{(k)}\right|^2 & \text{for } d = v \\ -x_{\mathrm{u},v}^{(k)} \left|\boldsymbol{p}_d^{(k),\mathrm{H}}\boldsymbol{h}_{\mathrm{u},v}^{(k)}\right|^2 & \text{for } d > v \\ 0 & \text{for } d < v \end{cases} \tag{3.141}$$

with the dual uplink noise covariance matrix

$$\boldsymbol{Q} = \mathrm{diag}(\boldsymbol{\mu}_{\mathrm{sum}} + \boldsymbol{\mu}_{\mathrm{mask}}^{(k)}) + \boldsymbol{H}_{\mathrm{u}}^{(k)}\mathrm{diag}(\boldsymbol{\mu}_{\mathrm{bmax}}^{(k)})\boldsymbol{H}_{\mathrm{u}}^{(k),\mathrm{H}}. \tag{3.142}$$

The transformation is derived from equating the downlink SNR

$$SNR_l^{(k)} = \frac{x_l^{(k)}|\boldsymbol{h}_l^{(k),\mathrm{T}}\boldsymbol{p}_l^{(k)}|^2}{1 + \sum_{d>l} x_d^{(k)} \left|\boldsymbol{h}_l^{(k),\mathrm{T}}\boldsymbol{p}_d^{(k)}\right|^2} \tag{3.143}$$

with the dual uplink SNR

$$SNR_{\mathrm{u},l}^{(k)} = \frac{x_{\mathrm{u},l}^{(k)}|\boldsymbol{h}_l^{(k),\mathrm{T}}\boldsymbol{p}_l^{(k)}|^2}{\boldsymbol{p}_l^{(k),\mathrm{H}}\boldsymbol{Q}^{(k)}\boldsymbol{p}_l^{(k)} + \sum_{v<l} x_{\mathrm{u},v}^{(k)} \left|\boldsymbol{h}_v^{(k),\mathrm{T}}\boldsymbol{p}_l^{(k)}\right|^2}. \tag{3.144}$$

In [55], projected gradient steps are proposed to update the Lagrange multipliers. For the optimization problem according to Eq. (3.133), the projected gradient steps are Eq. (3.103) for the per-line sum-power constraint, Eq. (3.101) for the spectral mask constraint and (3.104) for the bit loading upper bound.

Algorithm 8 summarizes the capacity calculation algorithm derived from [55], extended by the bit loading upper bound. While capacity is not achievable in an implementable system, it is an important tool to analyze performance losses caused by certain implementation limitations, as shown in the analysis from [1].

Three cases are of interest,

- the unbounded channel capacity without bit loading limitation, giving the theoretical capacity of the G.fast channel,
- the bounded capacity with bit loading constraint, showing the theoretical performance considering analog front-end limitations and

Algorithm 8 Capacity of the G.fast channel with per-line sum-power and per-line spectral mask constraint

Initialize $\mu_{\text{sum}}, \mu_{\text{mask}}^{(k)}, \mu_{\text{bmax}}^{(k)}$
repeat
  for $k = 1$ **to** K **do**
    repeat
      Do fixed-point iteration of Eqs. (3.136), (3.138)
    until Convergence
    Calculate uplink equalizer Eq. (3.135)
    Derive downlink precoder Eq. (3.139)
    Calculate transform matrix Eq. (3.141)
    Perform US-DS transform of $\boldsymbol{x}^{(k)}$ (3.140)
    Perform Lagrangian update Eq. (3.101)
    Perform Lagrangian update Eq. (3.104)
  end for
  Perform Lagrangian update Eq. (3.103)
until Convergence

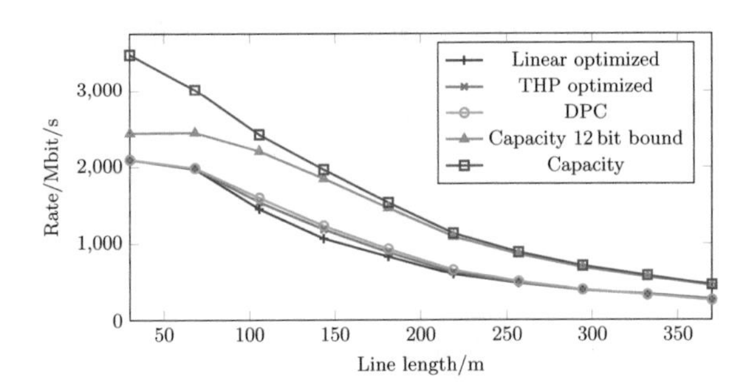

Fig. 3.16 Rate versus reach curves for channel capacity and implementation losses on a 30 pair DTAG PE06 binder with non-co-located CPEs

- the achievable rate with dirty paper coding and the G.fast specific losses , indicated by the SNR gap Γ and the coding overhead.

Comparing the mentioned capacity results with the precoding strategies proposed in this chapter gives an indication of the quality of the proposed algorithms. Figure 3.16 shows the rate versus reach curves for unbounded channel capacity, the impact of the 12 bit limitation on capacity and the rates achieved by dirty paper coding in combination with the actual coding and modulation scheme used in G.fast. The results are discussed in more detail in Sect. 3.7.

3.6 Spectrum Optimization with Protection of Legacy Services

As mentioned in Sect. 1.1, different digital subscriber line technologies coexist within the network. A new technology such as G.fast is introduced gradually and coexists with the legacy technologies.

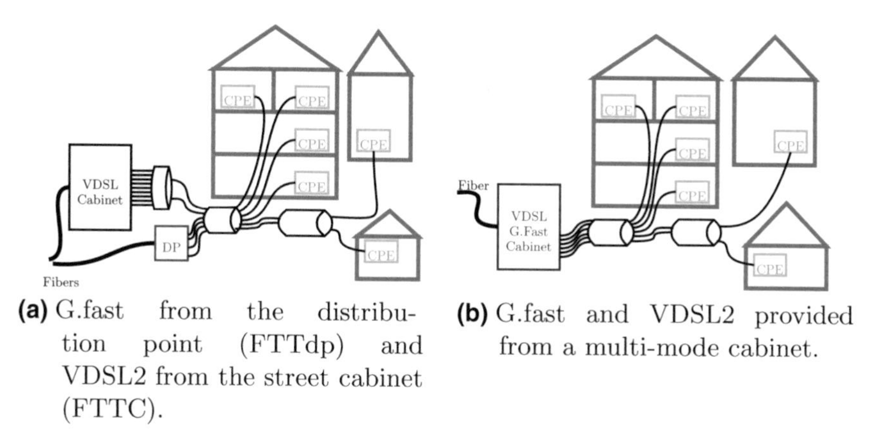

(a) G.fast from the distribution point (FTTdp) and VDSL2 from the street cabinet (FTTC).

(b) G.fast and VDSL2 provided from a multi-mode cabinet.

Fig. 3.17 Two network architectures with G.fast coexisting with VDSL2

In most migration scenarios, the VDSL2 FTTC architecture is extended [56] by the FTTdp architecture using G.fast [3, 57]. Some of the subscribers are still served with the legacy service, while others have been upgraded to G.fast in the same cable bundle.

Two possible architectures of G.fast coexisting with VDSL2 are shown in Fig. 3.17. In Fig. 3.17a, some lines are served by the VDSL2 cabinet, while the G.fast DPU is placed at a different location, closer to the subscribers.

Figure 3.17b shows another scenario of a multi-mode street cabinet, serving VDSL2 and G.fast in the FTTC architecture. In this case, VDSL2 and G.fast signals come from the same node. The node may also be a multi-mode DPU in a FTTdp architecture, supplying some legacy subscribers with a VDSL2 service.

Both G.fast deployment strategies result in mutual couplings between G.fast and VDSL2 services, referred to as alien crosstalk.

3.6.1 *Comparison of Legacy Service and G.fast*

In the coexistence cases of Fig. 3.17, a frequency division duplexing system (VDSL2) interferes with a time division duplexing system (G.fast). In this case, near-end crosstalk (NEXT) is present where the upstream of one service disturbs the downstream transmission of the other service and vice versa.

To analyze NEXT, upstream and downstream transmission are analyzed together. Therefore, downstream signals are denoted with subscript $(.)_{\mathrm{ds}}$ and upstream signals with $(.)_{\mathrm{us}}$. VDSL signals are denoted as $(.)_{\mathrm{V}}$ and G.fast signals as $(.)_{\mathrm{F}}$.

G.fast as well as VDSL2 are based on DMT modulation with K carriers on L lines in upstream (us) and downstream (ds) direction, but use different subcarrier spacings Δf_l. The transmit power $p_{\mathrm{ds},l}^{(k)}$ for downstream and $p_{\mathrm{us},l}^{(k)}$ for upstream can be configured for each line $l = 1, \ldots, L$ and carrier $k = 1, \ldots, K$ to achieve the desired power spectral densities (PSD) $\psi_{\text{ds},l}(f)$ and $\psi_{\text{us},l}(f)$ according to

$$p_{(\text{ds/us}),l}^{(k)} = \int_{f_k - \Delta f_l/2}^{f_k + \Delta f_l/2} \psi_{(\text{ds/us}),l}(f) df \tag{3.145}$$

with a subcarrier spacing Δf_l and a carrier frequency $f_k = k \Delta f_l$. The line indices are grouped into lines using VDSL2, $l \in \mathbb{I}_{\text{V}}$, and G.fast, $l \in \mathbb{I}_{\text{F}}$. VDSL2 and G.fast use different subcarrier spacings, $\Delta f_l = 4.3125$ kHz for $l \in \mathbb{I}_{\text{V}}$ and $\Delta f_l = 51.75$ kHz for $l \in \mathbb{I}_{\text{F}}$.

The spectral characteristics of both services are very different. G.fast uses a low per-line sum transmit power of $p_{\text{sum},l} = 4$ dBm for $l \in \mathbb{I}_{\text{F}}$, which is spread over a wide frequency band from 2.2 to 106 MHz or 212 MHz, while VDSL2 uses more than 10 times higher transmit power of $p_{\text{sum},l} = 14.5$ dBm for $l \in \mathbb{I}_{\text{V}}$ which is concentrated on a smaller frequency band from 25 KHz to 17.6 MHz. VDSL2 runs a DMT symbol rate of $1/t_{\text{sym},l} = 4$ kHz for $l \in \mathbb{I}_{\text{V}}$ while G.fast uses a much higher symbol rate of $1/t_{\text{sym}\,l} = 48$ kHz for $l \in \mathbb{I}_{\text{F}}$.

3.6.2 Mutual Couplings Between G.fast and VDSL2

Due to the different duplexing schemes and the overlapping spectrum between 2.2 and 17.6 MHz, there is crosstalk between G.fast and VDSL2. As shown in Fig. 3.18, there are four different coupling paths.

Each of the four receiving points (VDSL2 Cabinet, VDSL2 CPE, G.fast DP, G.fast CPE) experiences noise from three sources. For the affected (victim) line v, it is caused by NEXT (near end crosstalk) $\psi_{\text{NEXT},vd}(f)$ and far-end crosstalk (FEXT) $\psi_{\text{FEXT},vd}(f)$ from the disturber line d and receiver noise ψ_{n}.

Self-FEXT, the far-end crosstalk between lines of the same service is reduced by crosstalk cancelation. Therefore, G.fast victim lines $v \in \mathbb{I}_{\text{F}}$ are disturbed only by VDSL2 lines $d \in \mathbb{I}_{\text{V}}$ and the vice versa. The corresponding crosstalk transfer functions are downstream FEXT $H_{\text{FEXT ds},vd}^{(k)}$ and upstream FEXT $H_{\text{FEXT us},vd}^{(k)}$ and DP-side NEXT $H_{\text{NEXT dp},vd}^{(k)}$ and CPE-side NEXT $H_{\text{NEXT cpe},vd}^{(k)}$ for subcarrier k.

The interference plus noise PSD $\psi_{\text{in},v}(f)$ is given by

$$\psi_{\substack{\text{in}\\(\text{ds/us})},v}(f) = \sum_{d \in \mathbb{I}_{\text{dist},v}} \left(\psi_{\substack{\text{FEXT}\\(\text{ds/us})},vd}(f) + \psi_{\substack{\text{NEXT}\\(\text{cpe/dp})},vd}(f) \right) + \psi_{\text{n}} \tag{3.146}$$

where $\mathbb{I}_{\text{dist},v}$ are the indices of lines interfering v. The background noise is assumed to be additive white Gaussian (AWGN) zero-mean noise with a flat noise PSD ψ_{n}. The NEXT and FEXT PSDs used in Eq. (3.146) are derived by multiplication of the disturber transmit PSD $\psi_d(f)$ with the squared crosstalk transfer function, e. g.,

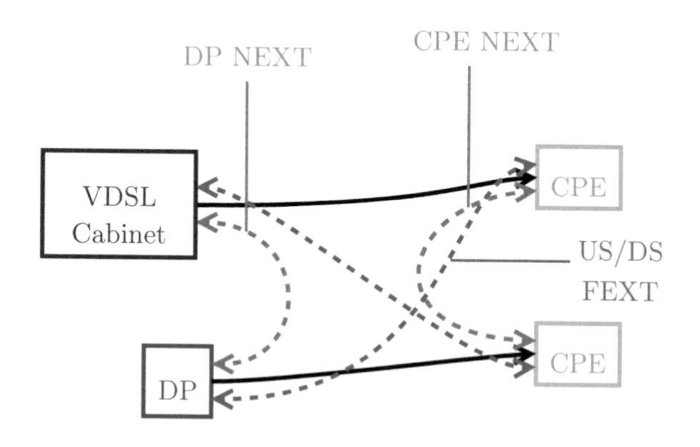

Fig. 3.18 Crosstalk couplings between G.fast and VDSL2

$$\psi_{\text{NEXT dp},vd}(f) = \psi_{\text{ds}\,d}(f)|H_{\text{NEXT dp},vd}(f)|^2 \tag{3.147}$$

for the DP-side NEXT $H_{\text{NEXT dp},vd}(f)$.[12] The crosstalk from disturber d to victim v is modeled according to Chap. 2.

The interference plus noise PSD can be converted into a noise power per tone according to $p_{\text{in},v}^{(k)} = \int_{f_k-\Delta f/2}^{f_k+\Delta f/2} \psi_{\text{in},v}(f)df$, which gives the *SINR* to be

$$SINR_{(\text{ds/us}),v}^{(k)} = \frac{|H_{(\text{ds/us}),v}^{(k)}|^2 p_{(\text{ds/us}),v}^{(k)}}{p_{\text{in (ds/us)},v}^{(k)}} \tag{3.148}$$

with a direct channel gain $H_{\text{ds},v}^{(k)}$ in downstream and $H_{\text{us},v}^{(k)}$ in upstream direction for line v.

3.6.3 Coexistence Strategies

This analysis focuses on two coexistence strategies, crosstalk avoidance by using different frequency bands, as shown in Fig. 3.19a and overlapped spectrum, as shown in Fig. 3.19b. Alien crosstalk cancelation is analyzed as a method to improve performance in the overlapped spectrum case.

Crosstalk Avoidance Strategy

The straight-forward strategy to guarantee spectral compatibility between G.fast and VDSL2 is crosstalk avoidance (Fig. 3.19a). In this case, the VDSL2 frequency bands are completely excluded from the used G.fast frequency band and only the out-of-band spectrum overlaps. This approach results in a minimum disturbance of the legacy service, but it causes a substantial loss of the G.fast data rates, especially for

[12] Accordingly, the CPE-side NEXT is $\psi_{\text{NEXT cpe},vd}(f) = \psi_{\text{us},d}(f)|H_{\text{NEXT cpe},vd}(f)|^2$ and the FEXT PSDs are $\psi_{\text{FEXT (ds/us)},vd}(f) = \psi_{(\text{ds/us}),d}(f)|H_{\text{FEXT (ds/us)},vd}(f)|^2$.

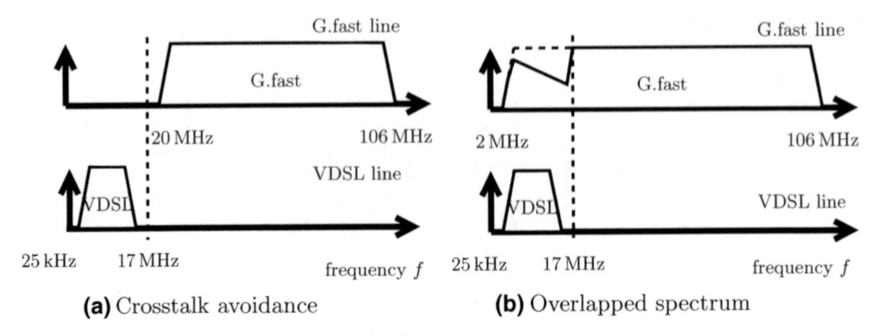

Fig. 3.19 Spectrum use for different G.fast/VDSL2 coexistence strategies

longer G.fast lines where the higher frequencies cannot be used. There is still some interference between both services due to their out-of-band transmit power (see e.g. Fig. 3.9).

Overlapped Spectrum Strategy

A less restrictive approach is to allow a spectral overlap between the services and optimize the transmit spectrum with spectral constraints. The optimization problem with per-line spectral mask and per-line sum-power constraints reads as

$$\max_{p_l^{(k)}} \sum_{l=1}^{L} R_{\text{ds},l} + R_{\text{us},l} \tag{3.149}$$
$$\text{s.t. } 0 \leq p_{(\text{ds/us}),l}^{(k)} \leq p_{\text{mask (ds/us)},l}^{(k)} \; \forall k = 1, \ldots, K; \; l = 1, \ldots, L$$
$$\text{s.t. } \sum_{k=1}^{K} p_{(\text{ds/us}),l}^{(k)} \leq p_{\text{sum},l} \; \forall l = 1, \ldots, L$$

with a sum-power limit $p_{\text{sum},l}$ and a spectral mask constraint $p_{\text{mask},l}^{(k)} = \int_{f_k-\Delta f/2}^{f_k+\Delta f/2} \psi_{\text{mask},l}(f)df$ as introduced in Sect. 3.1.6.

No coordination between legacy lines and G.fast lines is assumed. Each group of lines determines the optimized transmit spectrum independent of the others. The systems are only coupled by the crosstalk which each of the receivers observes. In this case, each of the services performs spectrum optimization according to Sect. 3.1.6 or according to Sects. 3.2 and 3.3. This corresponds to an iterative water-filling procedure, and converges to a locally optimal solution [3].

Sometimes, sum-rate optimization is not sufficient, because it is required to protect the legacy service. This is done by limiting the crosstalk into VDSL2 to a maximum value p_{nmax} according to

$$p_{\text{in (ds/us)},v}^{(k)} \leq p_{\text{nmax}} \tag{3.150}$$

for all G.fast lines $d \in \mathbb{I}_{\text{F}}$ disturbing legacy line $v \in \mathbb{I}_{\text{V}}$. A similar method is used in VDSL2 to protect ADSL lines or in non-vectored VDSL2 in upstream direction, which is called power back-off (PBO) [10]. Solving Eq. (3.149) including the constraint (3.150) is not feasible in practice, as it requires knowledge of the crosstalk between G.fast and VDSL2 lines as well as a central coordination for both services.[13]

With the help of crosstalk statistics, more precisely, with worst case values for aggregate DP-side NEXT $h^{(k)}_{\text{NEXTsum dp}}$, aggregate CPE-side NEXT $h^{(k)}_{\text{NEXTsum cpe}}$ and FEXT power sum $h^{(k)}_{\text{FEXTsum}}$, a more strict limit PSD for the G.fast lines is defined, which guarantees that (3.150) is satisfied with a certain probability.

Approximations for worst case cabinet NEXT and FEXT have been proposed in [58] for frequencies up to 30 MHz, which is sufficient for the alien crosstalk analysis. The channel model of Chap. 2 indicates that the worst case approximations presented there can be applied with small changes[14] for this application. Therefore, the following approximations are used for the worst case couplings

$$|h_{\text{NEXTsum dp}}(f)|^2 = N_{\text{dist}}^{0.6} \left(\frac{f}{f_0}\right)^{1.5} 10^{-44/10} \tag{3.151}$$

$$|h_{\text{NEXTsum cpe}}(f)|^2 = N_{\text{dist}}^{0.6} \left(\frac{f}{f_0}\right)^{0.75} 10^{-44/10} \tag{3.152}$$

$$|h_{\text{FEXTsum}}(f)|^2 = N_{\text{dist}}^{0.6} \left(\frac{f}{f_0}\right) \frac{d_{\text{avg}}}{1\text{km}} 10^{-39/10} \tag{3.153}$$

with $f_0 = 1$ MHz and a number of N_{dist} disturbers and the average disturber line length d_{avg}. Figure 3.20 shows the comparison between these approximations and the crosstalk model according to Chap. 2.

VDSL2 uses different frequency bands for upstream and downstream direction, which must be considered to derive the power back-off PSD mask. The frequencies used for VDSL2 downstream transmission are $f \in \mathcal{F}_{\text{Vds}}$ and the frequencies used for upstream transmission are $f \in \mathcal{F}_{\text{Vus}}$. It is assumed that the information abut the upstream and downstream bands is known to derive the G.fast PSD masks. The actual interference power from VDSL2 into G.fast also depends on the upstream/downstream ratio of the G.fast lines. But the ratio is not used to calculate the power back-off masks, which gives the worst case.

The power limit $\hat{p}^{(k)}_{\text{mask (ds/us)},l}$ including the power back-off (PBO) for G.fast lines for Eq. (3.149) is derived from the original PSD limit $p^{(k)}_{\text{mask (ds/us)},l}$ by

$$\hat{p}^{(k)}_{\substack{\text{mask}\\ \text{(ds/us)}},l} = \begin{cases} \min\left(p^{(k)}_{\substack{\text{pbo}\\ \text{(ds/us)}}}, p^{(k)}_{\substack{\text{mask}\\ \text{(ds/us)}},l}\right) & \forall k : f_k \in \mathcal{F}_{\text{Vus}} \cup \mathcal{F}_{\text{Vds}} \\ & \forall l \in \mathbb{I}_{\text{F}} \\ p^{(k)}_{\text{mask (ds/us)},l} & \text{otherwise.} \end{cases} \tag{3.154}$$

---

[13]Usually, the equipment in the DP and the street cabinet are owned by different network operators which means that they are not willing to cooperate, even if it would be technically possible.

[14]The scaling factors in Eqs. (3.151)–(3.153) are increased by 6 dB with respect to the definitions in [58] and the model for CPE-side NEXT is added in Eq. (3.152).

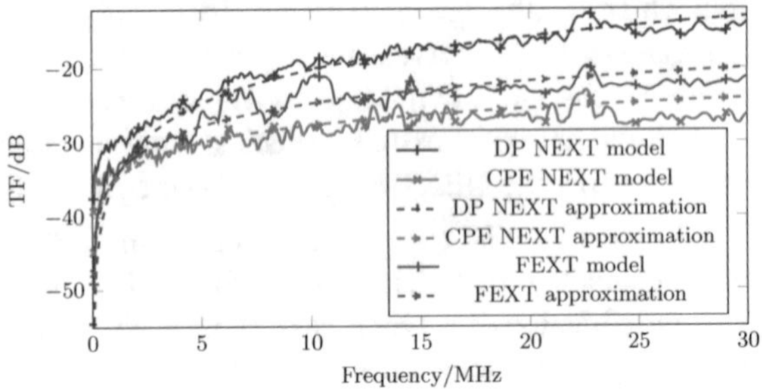

Fig. 3.20 Worst case NEXT and FEXT for 0.5 mm PE line, derived from the channel model according to Chap. 2 (solid lines) versus the worst case approximations according to Eq. (3.153) (dashed lines) for VDSL2 frequencies up to 30 MHz

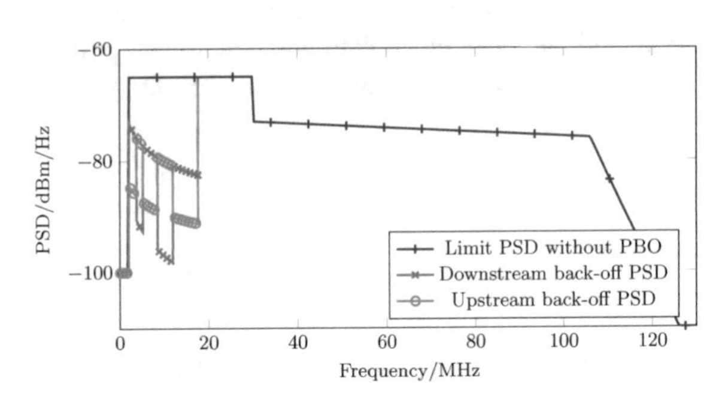

Fig. 3.21 The G.fast 106 MHz limit PSD (blue) in comparison with the downstream and upstream back-off PSDs for G.fast with VDSL2 protection for a scenario of co-located G.fast DP and VDSL2 cabinet (Fig. 3.17b)

The upstream and downstream power constraint values $p^{(k)}_{\text{pbo ds}}$ and $p^{(k)}_{\text{pbo us}}$ are derived from upstream and downstream back-off PSD masks for G.fast $\psi_{\text{pbo ds}}(f)$ and $\psi_{\text{pbo us}}(f)$ using Eq. (3.145) and the masks are given by

$$\psi_{\text{pbo ds}}(f) = \begin{cases} \dfrac{\psi_{\text{nmax}}}{|h_{\text{NEXTsum dp}}(f)|^2 |h_{\text{dp cab}}(f)|^2} & \text{for } f \in \mathcal{F}_{\text{Vus}} \\ \dfrac{\psi_{\text{nmax}}}{|h_{\text{FEXTsum}}(f)|^2} & \text{for } f \in \mathcal{F}_{\text{Vds}} \end{cases} \tag{3.155}$$

$$\psi_{\text{pbo us}}(f) = \begin{cases} \dfrac{\psi_{\text{nmax}}}{|h_{\text{FEXTsum}}(f)|^2 |h_{\text{dp cab}}(f)|^2} & \text{for } f \in \mathcal{F}_{\text{Vus}} \\ \dfrac{\psi_{\text{nmax}}}{|h_{\text{NEXTsum cpe}}(f)|^2} & \text{for } f \in \mathcal{F}_{\text{Vds}} \end{cases} \tag{3.156}$$

where $h_{\text{dp cab}}(f)$ represents the transfer function of the cable between DP and street cabinet and $\psi_{\text{nmax}} = p_{\text{nmax}}/\Delta f$. The resulting back-off PSDs for a multi-mode cabinet are shown in Fig. 3.21.

Alien Crosstalk Cancelation

In some cases, a sense line for alien crosstalk estimation is available as well as a second receiver path to detect the crosstalk signal. This allows to reduce alien crosstalk at the receiver side to some degree. Due to the fact that there is no alignment between the alien crosstalk and the desired signal, adaptive time domain filters are

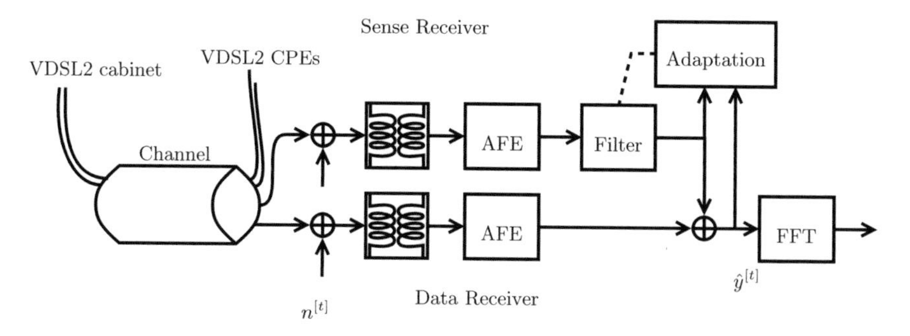

Fig. 3.22 Receiver with alien crosstalk cancelation

the most convenient way to remove crosstalk, which gives the receiver block diagram as shown in Fig. 3.22.

Assuming two receive signals $\hat{y}^{[t]}$ for the data line and $\hat{y}^{[t]}_{\text{sense}}$ for the sense line, the signal $\hat{y}^{[t]}_{\text{out}}$, which is the data line receive signal where crosstalk is canceled, the setup is described as

$$\hat{y}^{[t]}_{\text{out}} = \hat{y}^{[t]} - \sum_{i=1}^{N_\text{f}} \hat{y}^{[t+1-i]}_{\text{sense}} h^{[i]}_{\text{filter}}. \tag{3.157}$$

The cancelation filter vector is $\boldsymbol{h}_{\text{filter}}$ with a filter length N_f. The equalizer filter $\boldsymbol{h}_{\text{filter}}$ can be trained to minimize the mean squared error at the filter output, i.e. the difference between the actual receive signal and the alien interference-free receive signal. The desired signal of the G.fast line in time domain can be obtained during the quiet times of uplink and downlink, because the desired signal is zero at that time.

The filter update is given by

$$h^{[t+1]}_{\text{filter},i} = h^{[t]}_{\text{filter},i} - \alpha_{\text{filter}} e^{[t]} y^{[t-i+1]}_{\text{sense}} \tag{3.158}$$

with the update step size α_{filter}.

The alien crosstalk cancelation filter requires a second receive path as well as the adaptive canceler filter $\boldsymbol{h}_{\text{filter}}$. But the additional crosstalk cancelation allows to reduce the influence of alien crosstalk for deployment cases, where the additional sense line, as shown in Fig. 3.22, is available.

An alternative lower complexity approach to alien crosstalk cancelation using frequency domain filtering is shown in Appendix B.13. Frequency domain alien crosstalk cancelation is not able to compensate for intersymbol interference effects of alien crosstalk.

Simulation results for alien crosstalk cancelation in a specific scenario with potential benefits are shown in Appendix B.14. In general, the following conditions have to be satisfied to benefit from alien crosstalk cancelation.

- The crosstalk observed on the sense line and the data line is correlated.
- The crosstalk on the sense line is sufficiently strong to avoid noise enhancement through the canceler filter.

These conditions are satisfied when the sense line is one line within the cable binder and when there is only a single alien disturber or a small number of alien disturbers.

The minimum mean squared error criteria used to optimize the alien crosstalk canceler gives the optimal trade-off between crosstalk cancelation and noise enhancement in the mean squared error sense. In general, alien crosstalk cancelation can be used as an additional step to improve the performance of G.fast systems coexisting with VDSL2 and using overlapped spectrum methods.

3.7 Discussion

On the G.fast high frequency channel as discussed in Chap. 2, optimized crosstalk mitigation and spectrum allocation schemes are a critical component to achieve a high performance and quality of service. The studies of existing precoding and spectrum optimization methods for twisted pair communication in addition to the optimized zero-forcing precoding methods from Sects. 3.2.8 and 3.3.3 give a detailed picture of the performance of different precoding methods under realistic system assumptions.

Rate versus reach curves are used to compare different precoding and spectrum optimization methods. In contrast to wireless services, G.fast subscribers do not move and the line length as well as the channel quality corresponding to the line length does not change over time. Data rates as a function of line length are an important tool to compare different algorithms to verify that an improved transmission method is beneficial for all subscribers, those on long lines as well as those on short lines. Appendix B.15 gives more details on the topic.

As the precoding simulations are performed on cable bundles including many subscribers, there are two natural ways of creating the rate versus reach curve. Cable binders with all CPEs ending at the same length as in Fig. 3.23a (co-located CPEs) and cable binders with non-co-located CPEs as in Fig. 3.23b where each line in the binder

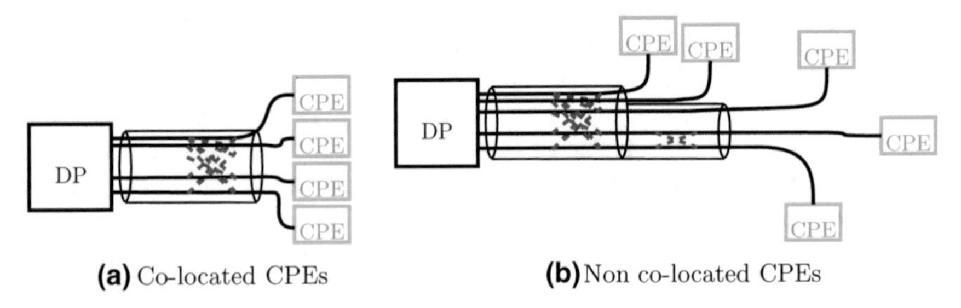

Fig. 3.23 Arrangement of CPEs for the co-located and non co-located network topology, considered in the simulation results

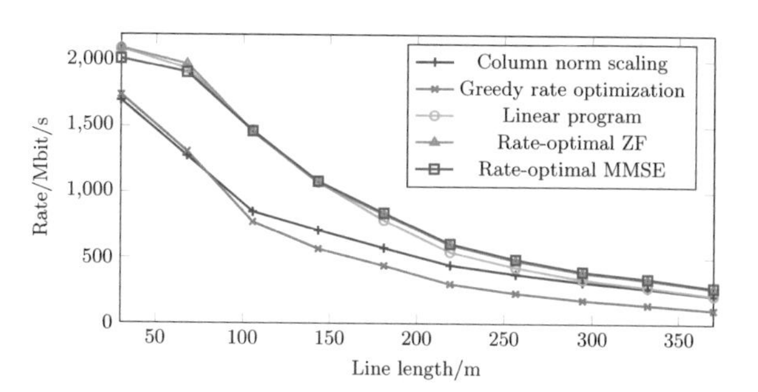

Fig. 3.24 Rate versus reach curves for comparison of different linear precoding and spectrum optimization methods on a 30 pair DTAG PE06 binder with non-co-located CPEs

has a different length. The latter one represents the actual network structure of the copper access network. More details can be found in the corresponding Appendices B.17–B.20.

With implementation complexity in mind, linear precoding methods play an important role and Fig. 3.24 shows the rate versus reach curves for the linear precoding methods investigated in this work. More details on the simulation conditions can be found in Appendix B.17.

The rate versus reach curves from Fig. 3.24 justify the use of convex optimization on G.fast precoding and spectrum optimization, because the optimized methods clearly outperform simpler approaches such as column norm scaling (Sect. 3.2.6) and the greedy bit loading algorithm according to Sect. 3.2.7.

Spectrum optimization by linear programming and especially the rate-optimal zero-forcing method described in Sects. 3.2.8 and 3.2.10 are the most promising methods for optimized linear precoding in G.fast, giving both, low complexity and high performance. Optimized zero-forcing methods operate very close to the upper bounds which are given by linear MMSE precoding (Sect. 3.2.9) for linear precoding and by dirty paper coding (Sect. 3.5) for THP. This is achieved by deactivating carriers with insufficient SNR to transmit at least one bit and it also results from the fact that G.fast operates at a high SNR range between 8 and 45 dB.

For G.fast, Tomlinson Harashima precoding is discussed intensively as an alternative precoding method (see e.g. [59–62]), adding some complexity to the precoding method for a potential performance advantage. Figure 3.25 shows the comparison between linear and non-linear zero-forcing precoding in the rate versus reach curve. More details on the simulation conditions can be found in Appendix B.18.

Figure 3.25 shows the performance advantage of THP over linear precoding at the medium loop length between 100 and 200 m to be up to 12% of the average rates. As for linear precoding, spectrum optimization is important to achieve the optimal data rates with zero-forcing THP. On long loops, the performance gain is not present due to the losses coming with THP, as discussed in Sect. 3.3.1. This results in linear precoding to have 3% higher data rates than THP and does not show TH precoding as a clearly superior precoding method.

The investigation is not complete without showing the capacity of the G.fast channel as an upper bound for any of the discussed transmission methods. Figure 3.26

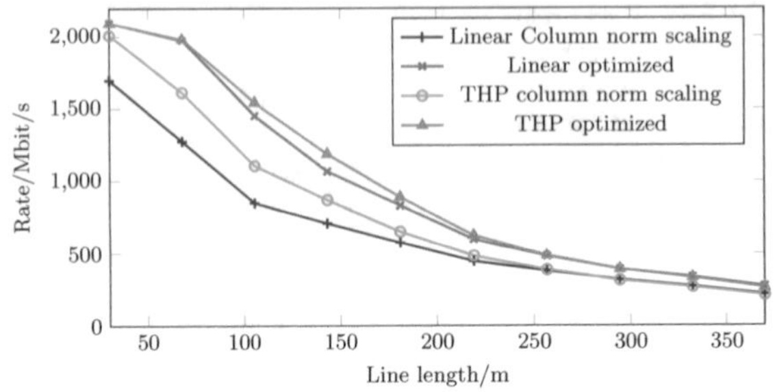

Fig. 3.25 Rate versus reach curves for linear and non-linear zero-forcing precoding on a 30 pair DTAG PE06 binder with non-co-located CPEs

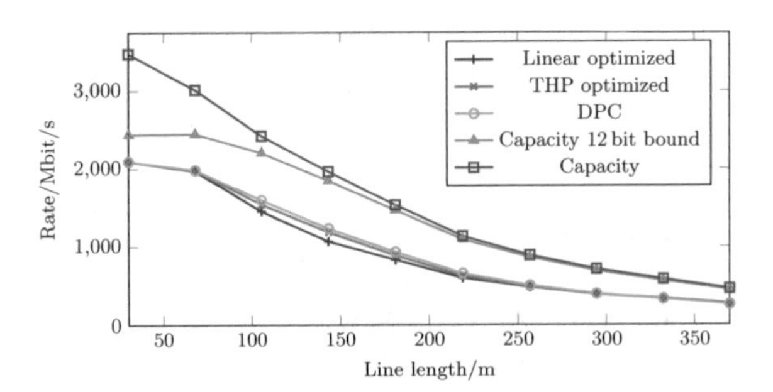

Fig. 3.26 Rate versus reach curves for channel capacity and implementation losses on a 30 pair DTAG PE06 binder with non-co-located CPEs

shows the rate versus reach curve resulting from a capacity calculation according to Sect. 3.5. More details can be found in Appendix B.20.

While dirty paper coding (DPC) [49] gives the upper bound for any precoding scheme for G.fast operating conditions, the capacity and the 12 bit bounded capacity provide the capacity limits of the channel for the given constraints. On short lines, the unbounded capacity is significantly higher than the capacity calculation with a 12 bit-constraint. However, the gap is small for the target line length between 100 and 200 m. On short loops, the constraint to 212 MHz bandwidth also limits the capacity, because the twisted pair channel would allow higher frequencies to be used.

Comparing Dirty Paper coding with implementable coding schemes shows that the presented optimization schemes bring the rate versus reach curve very close to the theoretical optimum. More room for improvement is indicated by the gap between DPC and the capacity curves. Reducing this gap requires to revisit the 6 dB SNR margin and the channel coding scheme, discussed in Sects. 3.1.4 and 3.1.4.

These results are obtained from channels which are used exclusively for G.fast. Simulation results for coexistence between G.fast and VDSL2 are discussed in Appendix B.16 showing that for the FTTdp architecture, coexistence can be maintained with small performance losses for legacy services and high data rates on the G.fast lines.

Figure 3.26 shows an upper bound on the achieved rates on G.fast channels. However, taking the rate allocation over time as an additional degree of freedom allows services to operate beyond the capacity at the sum-rate optimum, as it is discussed in the next chapter. Besides that, power consumption, as another important property of the G.fast system, is discussed.

References

1. Strobel, R., Joham, M., Utschick, W.: Achievable rates with implementation limitations for G.fast-based hybrid copper/fiber networks. In: IEEE International Conference on Communications (ICC) (2015)
2. Strobel, R., Barthelme, A., Utschick, W.: Zero-forcing and MMSE precoding for G.fast. In: IEEE Global Communications Conference (GLOBECOM) (2015)
3. Strobel, R., Utschick, W.: Coexistence of G.fast and VDSL in FTTdp and FTTC deployments. In: European Signal Processing Conference (EUSIPCO), pp. 1103–1107 (2015)
4. Weinstein, S., Ebert, P.: Data transmission by frequency-division multiplexing using the discrete fourier transform. IEEE Trans. Commun. Technol. **19**(5), 628–634 (1971)
5. ITU-T Rec. G.9701: Fast Access to Subscriber Terminals - Physical layer specification. ITU Recommendation (2015)
6. Wong, C.Y., Cheng, R.S., Lataief, K.B., Murch, R.D.: Multiuser OFDM with adaptive subcarrier, bit, and power allocation. IEEE J. Sel. Areas Commun. **17**(10), 1747–1758 (1999)
7. Strobel, R., Barthelme, A., Utschick, W.: Implementation aspects of nonlinear precoding for g.fast - coding and legacy receivers. In: European Signal Processing Conference (EUSIPCO) (2017)
8. ITU-T Rec. G.993.5-2010: Self-FEXT CANCELLATION (Vectoring) for Use with VDSL2 Transceivers (2010)
9. Oksman, V.: G.fast: proposal for signal constellations for G.fast (2013). ITU-T Contribution SG15/Q4a 2013-03-Q4-035
10. ITU-T Rec. G.993.2: Very high speed digital subscriber line transceivers 2 (VDSL2) (2006)
11. Sorbara, M.: G.fast: Constellations for use with Non-Linear Pre-Coding (2013). ITU-T Contribution SG15/Q4a 2013-01-Q4-028
12. Cioffi, J.: A multicarrier primer. ANSI T1E1 **4**, 91–157 (1991)
13. Kusume, K., Joham, M., Utschick, W., Bauch, G.: Cholesky factorization with symmetric permutation applied to detecting and precoding spatially multiplexed data streams. IEEE Trans. Signal Process. **55**(6), 3089–3103 (2007)
14. Yu, W., Varodayan, D.P., Cioffi, J.M.: Trellis and convolutional precoding for transmitter-based interference presubtraction. IEEE Trans. Commun. **53**(7), 1220–1230 (2005)
15. Neckebroek, J., Moeneclaey, M., Coomans, W., Guenach, M., Tsiaflakis, P., Moraes, R.B., Maes, J.: Novel bitloading algorithms for coded g. fast dsl transmission with linear and nonlinear precoding. In: IEEE International Conference on Communications (ICC), pp. 945–951. IEEE (2015)
16. Wei, L.F.: Trellis-coded modulation with multidimensional constellations. IEEE Trans. Inf. Theory **33**(4), 483–501 (1987)
17. Strobel, R., Oksman, V.: Dynamic Tone Ordering for SNR Margin Equalization (2017). US Patent App
18. ITU-T Rec. G.9700: Fast access to subscriber terminals (FAST) - Power spectral density specification. ITU Recommendation (2013)
19. Oksman, V., Strobel, R., Bry, C.: Dynamic update of transmission settings and robust management communication channel (2013). EU Patent App. EP2661009
20. Oksman, V., Strobel, R.: G.fast: proposal to define a synchronous multi-line OLR (2013). ITU-T Contribution SG15/Q4a 2013-05-Q4-063
21. Cendrillon, R., Moonen, M., Van den Bogaert, E., Ginis, G.: The linear zero-forcing crosstalk canceler is near-optimal in DSL channels. IEEE Global Commun. Conf. (GLOBECOM) **4**, 2334–2338 (2004)
22. Joham, M., Utschick, W., Nossek, J., et al.: Linear transmit processing in MIMO communications systems. IEEE Trans. Signal Process. **53**(8), 2700–2712 (2005)
23. Strobel, R., Oksman, V.: Channel estimation (2013). US Patent App. 14/062,983
24. Oksman, V., Strobel, R.: G.fast: Performance of channel estimation methods (2013). ITU-T Contribution SG15/Q4a COM 15 - C 2236 -E

25. Oksman, V., Strobel, R.: G.fast: impact of interpolation on convergence process during initialization (2013). ITU-T Contribution SG15/Q4a 2013-09-Q4-064
26. Maes, J., Nuzman, C.: Energy efficient discontinuous operation in vectored G.fast. In: IEEE International Conference on Communications (ICC), pp. 3854–3858. IEEE (2014)
27. Verbin, R.: G.fast: on the performance of a linear ZF precoder under high FEXT scenarios (2012). ITU-T Contribution SG15/Q4a 2012-11-4A-037
28. Campello, J.: Practical bit loading for DMT. In: IEEE International Conference on Communications (ICC), vol. 2, pp. 801–805. IEEE (1999)
29. Zanatta-Filho, D., Lopes, R.R., Ferrari, R., Suyama, R., Dortschy, B.: Bit Loading for Precoded DSL Systems. In: International Conference on Acoustics, Speech and Signal Processing (ICASSP), pp. 353–356 (2007)
30. Le Nir, V., Moonen, M., Verlinden, J.: Optimal power allocation under per-modem total power and spectral mask constraints in xdsl vector channels with alien crosstalk. In: International Conference on Acoustics, Speech and Signal Processing (ICASSP), vol. 3, pp. III–357. IEEE (2007)
31. Wiesel, A., Eldar, Y.C., Shamai, S.: Zero-forcing precoding and generalized inverses. IEEE Trans. Signal Process. **56**(9), 4409–4418 (2008)
32. Muller, F.C., Lu, C., Eriksson, P.E., Host, S., Klautau, A.: Optimizing power normalization for G. fast linear precoder by linear programming. In: IEEE International Conference on Communications (ICC), pp. 4160–4165. IEEE (2014)
33. Strobel, R., Smaoui, L., Oksman, V.: Rate-adaptive dynamic spectrum management (2013). US Patent App. 13/915,603
34. Yu, W., Lan, T.: Transmitter optimization for the multi-antenna downlink with per-antenna power constraints. IEEE Trans. Signal Process. **55**(6), 2646–2660 (2007)
35. Bogale, T.E., Vandendorpe, L.: Sum MSE optimization for downlink multiuser MIMO systems with per antenna power constraint: Downlink-uplink duality approach. In: International Symposium on Personal Indoor and Mobile Radio Communications (PIMRC), pp. 2035–2039. IEEE (2011)
36. Christensen, S.S., Agarwal, R., Carvalho, E., Cioffi, J.M.: Weighted sum-rate maximization using weighted MMSE for MIMO-BC beamforming design. IEEE Trans. Wirel. Commun. **7**(12), 4792–4799 (2008)
37. Fung, C.H.F., Yu, W., Lim, T.J.: Precoding for the multiantenna downlink: multiuser SNR gap and optimal user ordering. IEEE Trans. Commun. **55**(1), 188–197 (2007)
38. Gründinger, A., Joham, M., Gonzalez-Coma, J.P., Castedo, L., Utschick, W.: Average sum MSE minimization in the multi-user downlink with multiple power constraints. In: 48th Asilomar Conference on Signals, Systems and Computers, 2014, pp. 1279–1285. IEEE (2014)
39. Hunger, R., Utschick, W., Schmidt, D.A., Joham, M.: Alternating optimization for MMSE broadcast precoding. In: International Conference on Acoustics, Speech and Signal Processing (ICASSP), vol. 4, pp. IV–IV. IEEE (2006)
40. Oksman, V., Verbin, R., Strobel, R., Keren, R.: G.fast: channel estimation method for G.fast (2013). ITU-T Contribution SG15/Q4a 2013-01-Q4-039
41. Oksman, V., Strobel, R.: G.fast: performance of channel estimation methods up to 200 MHz (2012). ITU-T Contribution SG15/Q4a t 2012-11-4A-054
42. Oksman, V., , Strobel, R.: G.fast: on the channel estimation accuracy of LP and NLP (2016). ITU-T Contribution SG15/Q4a 2016-11-Q4-044
43. Widrow, B., Hoff, M.E., et al.: Adaptive Switching Circuits (1960)
44. Ginis, G., Cioffi, J.: Vectored-DMT: A FEXT canceling modulation scheme for coordinating users. In: IEEE International Conference on Communications (ICC), vol. 1, pp. 305–309. IEEE (2001)
45. Ginis, G., Cioffi, J.: Vectored transmission for digital subscriber line systems. IEEE J. Sel. Areas Commun. **20**(5), 1085–1104 (2002)
46. Tomlinson, M.: New automatic equaliser employing modulo arithmetic. Electron. Lett. **7**(5), 138–139 (1971)

47. Miyakawa, H., Harashima, H.: Information transmission rate in matched transmission systems with peak transmitting power limitation. In: Nat. Conf. Rec., Inst. Electron., Inform., Commun. Eng. of Japan, vol. 7, pp. 138–139 (1969)
48. Fischer, R.F., Windpassinger, C., Lampe, A., Huber, J.B.: Space-time transmission using Tomlinson-Harashima precoding. ITG Fachbericht pp. 139–148 (2002)
49. Costa, M.H.: Writing on dirty paper (corresp.). IEEE Trans. Inf. Theory **29**(3), 439–441 (1983)
50. Maes, J., Nuzman, C., Tsiaflakis, P.: Sensitivity of nonlinear precoding to imperfect channel state information in g. fast. In: European Signal Processing Conference (EUSIPCO), pp. 290–294. IEEE (2016)
51. Masouros, C., Sellathurai, M., Ratnarajah, T.: Interference optimization for transmit power reduction in Tomlinson–Harashima precoded MIMO downlinks. IEEE Trans. Signal Process. **60**(5), 2470–2481 (2012)
52. Emig, T.: Effects of Nonlinear Precoding on Viterbi Decoding and System Performance in Multi-user Copper Networks. Master's thesis, Technische Universität München (2016)
53. Barthelme, A., Strobel, R., Joham, M., Utschick, W.: Weighted MMSE Tomlinson–Harashima Precoding for G.fast. In: IEEE Global Communications Conference (GLOBECOM) (2016)
54. Yu, W.: Uplink-downlink duality via minimax duality. IEEE Trans. Inf. Theory **52**(2), 361–374 (2006)
55. Lanneer, W., Moonen, M., Tsiaflakis, P., Maes, J.: Linear and nonlinear precoding based dynamic spectrum management for downstream vectored g. fast transmission. In: IEEE Global Communications Conference (GLOBECOM). IEEE (2015)
56. Mazzenga, F., Petracca, M., Vatalaro, F., Giuliano, R., Ciccarella, G.: Coexistence of FTTC and FTTDp network architectures in different VDSL2 scenarios. In: Transactions on Emerging Telecommunications Technologies (2014)
57. Strobel, R.: Coexistence and the Migration to G.fast FTTdp (2015). G.fast Summit (2015). http://www.uppersideconferences.com/g.fast/-gfast2015_agenda_conference_day_1.html
58. ETSI TS 101 270-1: Transmission and Multiplexing (TM); Access transmission systems on metallic access cables; Very high speed Digital Subscriber Line (VDSL); Part 1: Functional requirements (2003)
59. Oksman, V., Strobel, R.: G.fast: Performance evaluation of linear and non-linear precoders (2013). ITU-T Contribution SG15/Q4a 2013-01-Q4-041R1
60. Oksman, V., Strobel, R.: G.fast: G.fast performance over KPN cable (2013). ITU-T Contribution SG15/Q4a COM 15 - C 0338 - E
61. Oksman, V., Strobel, R.: G.fast: simulation results for BT cable (2012). ITU-T Contribution SG15/Q4a 2012-11-4A-056
62. Oksman, V., Strobel, R.: G.fast: G.fast performance over Swisscom cable (2013). ITU-T Contribution SG15/Q4a 2013-01-Q4-042

Chapter 4
Framing-Based Optimization

G.fast introduces a different duplexing scheme for DSL technologies. While previous DSL systems, e.g., ADSL/ADSL2 or VDSL2 use frequency division duplexing (FDD), G.fast uses time division duplexing (TDD). TDD has advantages in terms of power consumption because transmitter and receiver are not active at the same time and one of them can always be switched to low power mode. Besides that, the analog and mixed-signal design is simplified, because there is no echo signal present which couples from the transmitter into the receiver.

Besides the TDD scheme, G.fast introduces a method called "Discontinuous Operation" (DO) to further reduce the transmit and receive time within a frame, depending on the current data traffic [1]. The main reason to introduce DO is power consumption, but it can also be used to increase data rates.

Section 4.1 gives an introduction to the G.fast frame structure and the corresponding control mechanisms. The dependencies between discontinuous operation and crosstalk cancelation are investigated in Sect. 4.2.

Section 4.3 explains methods for power consumption minimization with rate constraints, which are based on DO. The optimal solution is compared to simplified implementable DO methods.

Section 4.4 demonstrates another use case of G.fast framing-based optimization, allowing maximization of peak data rates, which are temporarily available to individual subscribers.

4.1 G.fast Frame Structure

A TDD system requires a frame structure to maintain upstream and downstream times as well as time gaps to switch transmission direction. Besides that, the G.fast frame structure introduces dedicated symbols for channel estimation and overhead channels.

R. Strobel, *Channel Modeling and Physical Layer Optimization in Copper Line Networks*, Signals and Communication Technology,
https://doi.org/10.1007/978-3-319-91560-9_4

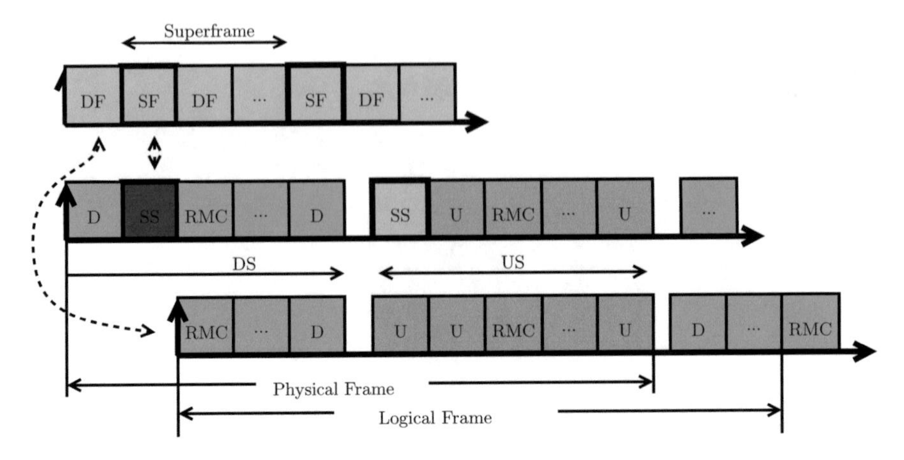

Fig. 4.1 Superframes and TDD frames in G.fast

4.1.1 TDD Frame Format

G.fast transmission is organized in superframes and TDD frames, as shown in Fig. 4.1. A superframe consists of 8 or 12 TDD frames with an overall length of 6 ms.[1] The first TDD frame of the superframe is the sync frame (SF), which includes sync symbols (SS) used for channel estimation in upstream and downstream direction. The data frames (DF) have the same number of symbols as the sync frames, but a data symbol on the sync symbol position.

There are two definitions of the TDD frame, as indicated in Fig. 4.1. The logical frame starts with the downstream robust management channel (RMC) symbol and ends with the last downstream symbol before the RMC symbol. The physical frame starts with the first symbol of the downstream period and ends with the last upstream symbol.

Each of the TDD frames consists of a downstream and upstream period and $\frac{1}{2}$-symbol gaps between upstream and downstream periods to change transmission direction. The length of the gaps is adjusted with respect to the line delay such that there are no collisions between upstream and downstream at the DP-side or the CPE-side.

The downstream portion as well as the upstream portion of a TDD frame are further partitioned into a normal operation interval (NOI) and a discontinuous operation interval (DOI) as shown in Fig. 4.2 for the downstream portion.

NOI and DOI are used to implement discontinuous operation. During NOI, all links are active, i.e., the DPU transmitters or receivers must be enabled. During DOI, the lines may discontinue transmission, depending on the actual data rate and the

[1]The standard configuration uses $M_{\mathrm{SF}} = 8$ TDD frames in one superframe while the length of a TDD frame is $M_{\mathrm{F}} = 8$ DMT symbols. Alternatively, a superframe of $M_{\mathrm{SF}} = 12$ TDD frames with $M_{\mathrm{F}} = 23$ DMT symbols per TDD frame may be used.

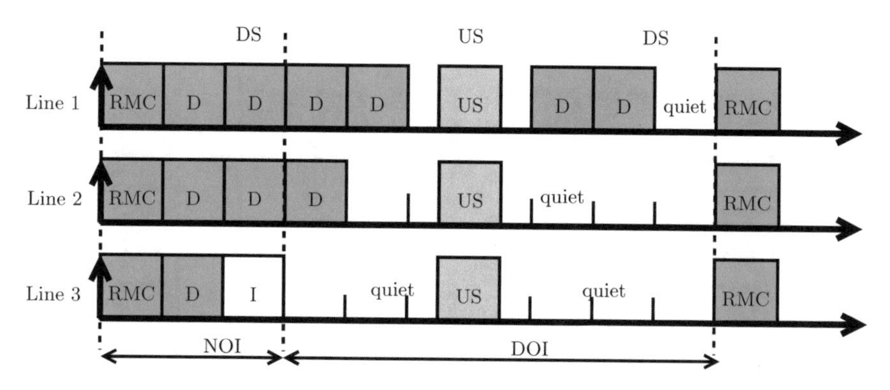

Fig. 4.2 NOI and DOI in G.fast

corresponding transmitters and receivers at the DPU may switch to low power mode until the end of the TDD frame.

Besides that, there are different symbol types as indicated in Fig. 4.2. The data (D) symbols carry only regular user data, while the RMC symbol carries the robust management channel and user data. It indicates the start of the logical frame and contains the framing information for the TDD frame, which is the number of symbols in the NOI and the number of transmitted data symbols as well as their position in the DOI [2, 3].

Furthermore, there are two types of symbols which do not carry data, idle (I) and quiet symbols. For idle symbols, the signal portion required for crosstalk cancelation is still transmitted on the corresponding lines. Quiet symbols have zero power at the precoder input as well as at the precoder output. The line driver and the analog front-end of the corresponding line are inactive. Quiet symbols are only allowed in the DOI.

Different transmitter and receiver settings are used in NOI and DOI. There is a NOI bit loading $\hat{b}^{(k)}_{\mathrm{NOI},l}$ and a DOI bit loading $\hat{b}^{(k)}_{\mathrm{DOI},l}$ as well as NOI gains $s^{(k)}_{\mathrm{NOI},l}$ and DOI gains $s^{(k)}_{\mathrm{DOI},l}$. These settings are precomputed and stored in the DPU and the CPEs in advance.

The stored settings reduce the communication overhead required to change between different configurations. In G.fast, only information about the transmit time of NOI and DOI is communicated in each TDD frame. The corresponding transmitter settings are exchanged at a much slower rate, e.g., by seamless rate adaptation as described in Sect. 3.1.7.

Figures 4.1 and 4.2 are specific for G.fast framing. At this point, a more general approach for TDD frame optimization is introduced, where the TDD frame is partitioned into configurations $t = 1, \ldots, T$. Each configuration change indicates a change of transmitter and receiver settings which is either a change of bit loading $\hat{b}^{(k),[t]}_l$ and gains $s^{(k),[t]}_l$ or a change of the active lines $\mathbb{I}^{[t]}_{\mathrm{a}}$.

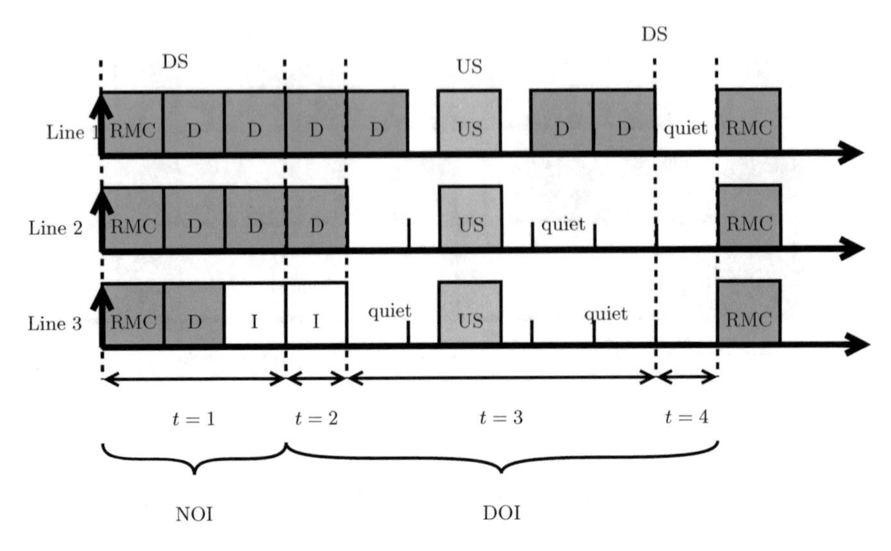

Fig. 4.3 TDD frame with four different configurations in downstream, $T = 4$

For each configuration t, there is a specific gain setting $\boldsymbol{S}^{(k),[t]}$, a bit allocation $\hat{b}^{(k),[t]}$, a precoder $\boldsymbol{P}^{(k),[t]}$ for downstream or an equalizer $\boldsymbol{G}^{(k),[t]}$ for upstream and a specific set of enabled lines $\mathbb{I}_{\mathrm{a}}^{[t]}$ [3–5].

The partitioning of the TDD frame in NOI and DOI is a special case of Fig. 4.3, e.g., the NOI is represented by $t = 1$ and the DOI is represented by $t = 2, \ldots, T$. The NOI bit allocation and gain tables are $\hat{b}_{\mathrm{NOI},l}^{(k)} = \hat{b}_l^{(k),[t=1]}$ and $s_{\mathrm{NOI},l}^{(k)} = s_l^{(k),[t=1]}$ while for all DOI configurations, there is only one bits and gains table, $\hat{b}_{\mathrm{DOI},l}^{(k)} = \hat{b}_l^{(k),[t>1]}$ and $s_{\mathrm{DOI},l}^{(k)} = s_l^{(k),[t>1]}$.

This framework is used to describe the dependencies between discontinuous operation and downstream precoding or upstream equalization and to introduce framing-based optimization of power consumption and data rates.

4.1.2 Data Rates in Discontinuous Operation

The data rates in discontinuous operation vary within the TDD frame due to the configuration-dependent bit allocation $\hat{b}_l^{(k),[t]}$. A rate vector $\boldsymbol{r}^{[t]} \in \mathbb{R}^L$ for each configuration t is introduced, containing the data rate that is achieved when a DMT symbol is transmitted with the settings of configuration t.

The instantaneous line rate for line l during configuration t is

$$R_l^{[t]} = \begin{cases} \frac{\eta}{t_{\mathrm{sym}}} \sum_{k=1}^{K} \hat{b}_l^{(k),[t]} - R_{\mathrm{oh}} & \text{for } l \in \mathbb{I}_{\mathrm{a}}^{[t]} \\ 0 & \text{otherwise,} \end{cases} \tag{4.1}$$

with symbol time t_{sym}, framing efficiency η and overhead channel rate R_{oh}, corresponding to Eq. (3.17) in Sect. 3.1.5. The rate vector for configuration t is then $\boldsymbol{r}^{[t]} = [R_1^{[t]}, \ldots, R_L^{[t]}]^{\text{T}}$. They are collected in a rate matrix $\boldsymbol{R} \in \mathbb{R}^{L \times T}$ according to

$$\boldsymbol{R} = \left[\boldsymbol{r}^{[1]}, \ldots, \boldsymbol{r}^{[T]}\right]. \tag{4.2}$$

The average data rate $R_{\text{avg},l}$ for line l depends on the frame timing, denoted by $\tau^{[t]}$, the time fraction of the TDD frame, where configuration t is used. The corresponding timing vector is $\boldsymbol{\tau} = [\tau^{[1]}, \ldots, \tau^{[T]}]$ where $0 \leq \tau^{[t]} \leq 1$ and $\sum_{t=1}^{T} \tau^{[t]} = 1$ holds.

The average rate vector $\boldsymbol{r}_{\text{avg}} = [R_{\text{avg},1}^{[t]}, \ldots, R_{\text{avg},L}^{[t]}]^{\text{T}}$ is then given by

$$\boldsymbol{r}_{\text{avg}} = \boldsymbol{R}\boldsymbol{\tau}. \tag{4.3}$$

4.2 Crosstalk Cancelation and Discontinuous Operation

The combination of precoding and discontinuous operation has been studied in [6, 7] for linear and nonlinear precoding and further investigated in [8] for linear precoding. [1, 4, 5] explain the combination of precoding and discontinuous operation for G.fast, including implementation limitations.

4.2.1 Idle and Quiet Symbols

The dependency between precoding and discontinuous operation can be explained with the difference between idle and quiet symbols. When linear or nonlinear precoding is used in downstream direction, the transmit signals at the precoder output are a combination of the input signals for all lines. Turning off one or more transmitters at the DPU will affect the receive signals at all receivers [7].

During a quiet symbol on line l, the DPU transmitter of the corresponding line is switched off. Herby, power is saved because because major parts of the analog and digital front-end, which are the main power consumers within the G.fast system besides the line driver, are switched off. The converter power consumption is independent of the transmit signal power, but they can be switched to low power mode during a quiet symbol.

For an idle symbol on line l, only the precoder input signal on line l is zero, while the transmitter remains enabled, as shown in Fig. 4.4a. For transmission of an idle symbol, no change of the precoder is required.

But during transmission of the quiet symbol, as shown in Fig. 4.4b, the secondary signal paths[2] are no longer available, which requires to change the precoder.

[2]The paths from the active lines modulators over the precoder into the discontinued line transmitter and from the discontinued line to the active lines receivers via the crosstalk.

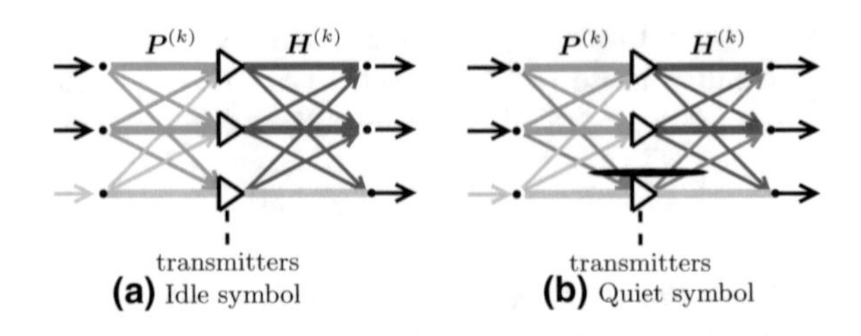

Fig. 4.4 Idle and quiet symbols in discontinuous operation (a) Idle symbol (b) Quiet symbol

Without loss of generality, the transmit and receive signal vectors are separated into an active lines signal vector $\boldsymbol{u}_{\mathrm{a}}(k)$ and a discontinued lines signal vector $\boldsymbol{u}_{\mathrm{d}}^{(k)} = \boldsymbol{0}$ such that the overall signal is $\boldsymbol{u}^{(k)} = [\boldsymbol{u}_{\mathrm{a}}^{(k),\mathrm{T}}, \boldsymbol{u}_{\mathrm{d}}^{(k),\mathrm{T}}]^{\mathrm{T}}$. The same holds for the precoder output signal vector $\boldsymbol{y}^{(k)} = [\boldsymbol{y}_{\mathrm{a}}^{(k),\mathrm{T}}, \boldsymbol{y}_{\mathrm{d}}^{(k),\mathrm{T}}]^{\mathrm{T}}$ and the receive signal vector $\hat{\boldsymbol{u}}^{(k)} = [\hat{\boldsymbol{u}}_{\mathrm{a}}^{(k),\mathrm{T}}, \hat{\boldsymbol{u}}_{\mathrm{d}}^{(k),\mathrm{T}}]^{\mathrm{T}}$.

4.2.2 Linear Precoding and Discontinuous Operation

For linear precoding, the block diagram of Fig. 3.11 in Sect. 3.2 is assumed. The channel $\boldsymbol{H}^{(k)}$, precoder $\boldsymbol{P}^{(k)}$ and equalizer $\boldsymbol{G}^{(k)}$ matrices are partitioned into four block matrices

$$\boldsymbol{P}^{(k)} = \begin{bmatrix} \boldsymbol{P}_{\mathrm{aa}}^{(k)} & \boldsymbol{P}_{\mathrm{ad}}^{(k)} \\ \boldsymbol{P}_{\mathrm{da}}^{(k)} & \boldsymbol{P}_{\mathrm{dd}}^{(k)} \end{bmatrix}, \quad \boldsymbol{H}^{(k)} = \begin{bmatrix} \boldsymbol{H}_{\mathrm{aa}}^{(k)} & \boldsymbol{H}_{\mathrm{ad}}^{(k)} \\ \boldsymbol{H}_{\mathrm{da}}^{(k)} & \boldsymbol{H}_{\mathrm{dd}}^{(k)} \end{bmatrix}, \tag{4.4}$$

$$\boldsymbol{G}^{(k)} = \begin{bmatrix} \boldsymbol{G}_{\mathrm{aa}}^{(k)} & \boldsymbol{G}_{\mathrm{ad}}^{(k)} \\ \boldsymbol{G}_{\mathrm{da}}^{(k)} & \boldsymbol{G}_{\mathrm{dd}}^{(k)} \end{bmatrix} \tag{4.5}$$

according to the active lines $\mathbb{I}_{\mathrm{a}}^{[t]}$ for the current configuration.

Assuming zero-forcing precoding,

$$\boldsymbol{G}^{(k)} \boldsymbol{H}^{(k)} \boldsymbol{P}^{(k)} = \boldsymbol{I} \tag{4.6}$$

holds when all lines are active. For any selection of discontinued lines during the discontinuous operation interval,

$$\boldsymbol{G}_{\mathrm{aa}}^{(k)} \boldsymbol{H}_{\mathrm{aa}}^{(k)} \boldsymbol{P}_{\mathrm{aa}}^{(k),\prime} = \boldsymbol{I} \tag{4.7}$$

shall hold.

This requires different precoding coefficients $\boldsymbol{P}_{\mathrm{aa}}^{(k),\prime} \neq \boldsymbol{P}_{\mathrm{aa}}^{(k)}$. Applying the matrix inversion lemma as explained in Appendix C.1, these coefficients are obtained by

$$\boldsymbol{P}_{\mathrm{aa}}^{(k),\prime} = \boldsymbol{P}_{\mathrm{aa}}^{(k)} - \boldsymbol{P}_{\mathrm{ad}}^{(k)} \boldsymbol{P}_{\mathrm{dd}}^{(k),-1} \boldsymbol{P}_{\mathrm{da}}^{(k)}. \tag{4.8}$$

With the new precoding coefficients, Eq. (4.7) is satisfied without changes of the diagonal equalizer coefficients $\boldsymbol{G}_{\mathrm{aa}}$ of the active lines at CPE side. As can be seen by Eq. (4.8), no additional information other than the precoder coefficients of all lines is needed to implement the correction. As the set of discontinued lines may change from symbol to symbol, this computation must be done at a very short time.

For zero-forcing precoding, the signal-to-noise ratio $SNR_l^{(k),[t]}$ can be derived from the equalizer coefficients and the transmit gain values. As the equalizer coefficients remain constant when Eq. (4.8) is used, the SNR can be determined by

$$SNR_l^{(k),[t]} = \frac{\left|s_l^{(k),[t]}\right|^2}{\left|g_l^{(k)}\right|^2 \sigma^2} \tag{4.9}$$

where only $s_l^{(k),[t]}$ changes for different configurations.

In case that interference allowing precoding methods are used or when approximations of Eq. (4.8) are used, there is residual crosstalk to be considered for the configuration-dependent SNR, which is given by

$$SNR_l^{(k),[t]} = \frac{\left|\boldsymbol{h}_l^{(k),\mathrm{T}} \boldsymbol{p}_l^{(k),[t]} s_l^{(k),[t]}\right|^2}{\left|g_l^{(k)}\right|^2 \left(\sigma^2 + \sum_{d \neq l} \left|\boldsymbol{h}_l^{(k),\mathrm{T}} \boldsymbol{p}_d^{(k),[t]} s_d^{(k),[t]}\right|^2\right)} \tag{4.10}$$

where the SNR depends on the gain scaling as well as the crosstalk due to the configuration-dependent precoder.

4.2.3 Nonlinear Precoding

The investigation of nonlinear precoding in combination with discontinuous operation is based on zero-forcing Tomlinson Harashima precoding as described in Sect. 3.3 and Fig. 3.14. The topic is discussed in [4–6].

The Tomlinson Harashima precoder requires two matrices, the feed-back matrix $\boldsymbol{P}_{\mathrm{b}}$ and the forward matrix $\boldsymbol{P}_{\mathrm{f}}$. The correction is applied to the forward matrix $\boldsymbol{P}_{\mathrm{f}}^{(k)}$ while the feedback matrix remains unchanged when lines discontinue. It is partitioned according to

$$\boldsymbol{P}_{\mathrm{f}}^{(k)} = \begin{bmatrix} \boldsymbol{P}_{\mathrm{f,aa}}^{(k)} & \boldsymbol{P}_{\mathrm{f,ad}}^{(k)} \\ \boldsymbol{P}_{\mathrm{f,da}}^{(k)} & \boldsymbol{P}_{\mathrm{f,dd}}^{(k)} \end{bmatrix}. \tag{4.11}$$

The update is given by [3]

$$\boldsymbol{P}_{\text{f,aa}}^{(k),\prime} = \boldsymbol{P}_{\text{f,aa}}^{(k)} + \boldsymbol{H}_{\text{aa}}^{(k),-1} \boldsymbol{H}_{\text{ad}}^{(k)} \boldsymbol{P}_{\text{f,da}}^{(k)}. \tag{4.12}$$

For the precoder update according to Eq. (4.12), it is required that the signal $\boldsymbol{u}_{\text{mod}}^{(k)}$ at the output of the nonlinear operation satisfies $u_{\text{mod},d}^{(k)} = 0$ for the discontinued lines $d \in \mathbb{I}_d$. This requires to select a specific encoding order. The line, which is discontinued first in the TDD frame is encoded last.

This is also the right strategy from a rate optimization point of view. The line which discontinued last is the one which requests the highest data rate and the line which is encoded first achieves the highest rate.

4.2.4 Spectral Constraints

The transmit power constraints, as described in Sect. 3.2.2, shall be satisfied at any time when lines are discontinued. According to the definition of the power constraints in [9], the transmit power constraints shall be satisfied for the time the transmission is active, i.e., the quiet times do not count to the average transmit power. Therefore, the transmit gains $\boldsymbol{S}^{(k),[t]}$ must be changed with respect to the precoder coefficients $\boldsymbol{P}^{(k),[t]}$ for each configuration t.

The per-line spectral mask constraint then reads as

$$\operatorname{diag}\left(\boldsymbol{P}^{(k),[t]} \boldsymbol{S}^{(k),[t]} \boldsymbol{S}^{(k),[t],\text{H}} \boldsymbol{P}^{(k),[t],\text{H}}\right) \leq \boldsymbol{p}_{\text{mask}}^{(k)} \quad \forall k = 1, \dots, K \quad \forall t = 1, \dots, T \tag{4.13}$$

and the per-line sum-power constraint is

$$\sum_{k=1}^{K} \left(\boldsymbol{P}^{(k),[t]} \boldsymbol{S}^{(k),[t]} \boldsymbol{S}^{(k),[t],\text{H}} \boldsymbol{P}^{(k),[t],\text{H}}\right) \leq \boldsymbol{p}_{\text{sum}} \quad \forall t = 1, \dots, T. \tag{4.14}$$

The amount of data to be communicated per configuration is large, e.g., gain values and bit allocation values for each carrier and line. Therefore, the gain and bit allocation values are exchanged in advance and stored at the transmitter as well as at the receiver. From an implementation perspective, the number of different configurations must be limited.

---

[3] The effective channel for the active lines $\boldsymbol{H}_{\text{eff,aa}}^{(k)} = \boldsymbol{H}_{\text{aa}}^{(k)} \boldsymbol{P}_{\text{f,aa}}^{(k)} + \boldsymbol{H}_{\text{ad}}^{(k)} \boldsymbol{P}_{\text{f,da}}^{(k)}$ is kept constant between NOI and DOI by changing the precoder coefficients for the active lines according to $\boldsymbol{H}_{\text{eff,aa}}^{(k)} = \boldsymbol{H}_{\text{aa}}^{(k)} \left(\boldsymbol{P}_{\text{f,aa}}^{(k)} + \boldsymbol{H}_{\text{aa}}^{(k),-1} \boldsymbol{H}_{\text{ad}}^{(k)} \boldsymbol{P}_{\text{f,da}}^{(k)}\right)$.

G.fast allows only two different settings, one for the NOI configuration and one for all DOI configurations. Therefore, the NOI constraint set, $t = 1$, set still read as Eqs. (4.13) and (4.14), but for the DOI, $t = 2, \ldots, T$, the constraints change to

$$\operatorname{diag}\left(\boldsymbol{P}_{\mathrm{aa}}^{\prime(k),[t]}\boldsymbol{S}_{\mathrm{DOI}}^{(k)}\boldsymbol{S}_{\mathrm{DOI}}^{(k),\mathrm{H}}\boldsymbol{P}_{\mathrm{aa}}^{\prime(k),[t],\mathrm{H}}\right) \leq \boldsymbol{p}_{\mathrm{mask}}$$
$$\forall k = 1, \ldots, K \;\; \forall t = 2, \ldots, T, \tag{4.15}$$

$$\sum_{k=1}^{K}\operatorname{diag}\left(\boldsymbol{P}_{\mathrm{aa}}^{\prime(k),[t]}\boldsymbol{S}_{\mathrm{DOI}}^{(k)}\boldsymbol{S}_{\mathrm{DOI}}^{(k),\mathrm{H}}\boldsymbol{P}_{\mathrm{aa}}^{\prime(k),[t],\mathrm{H}}\right) \leq \boldsymbol{p}_{\mathrm{sum}}$$
$$\forall t = 2, \ldots, T. \tag{4.16}$$

The G.fast-specific constraints according to Eqs. (4.15) and (4.16) are more strict than the general constraints of Eqs. (4.13) and (4.14), because of a single gain matrix $\boldsymbol{S}_{\mathrm{DOI}}^{(k)}$ which is required to satisfy the power constraints for multiple configurations $t = 2, \ldots, T$. This results in lower DOI rates for the G.fast case.

4.2.5 Equalization and Discontinuous Operation

In the G.fast uplink direction, there is no difference between idle and quiet symbols from a transmitter perspective. The transmitter operation and the transmit power constraints remain unchanged.

However, switching off receiver components such as analog-to-digital conversion requires to recompute the uplink equalizer matrices. The analog-to-digital conversion at the receiver is a major part of the overall power consumption.

The update rule is derived from the zero-forcing assumption $\boldsymbol{G}^{(k)} = \left(\boldsymbol{H}^{(k)}\boldsymbol{S}^{(k)}\right)^{-1}$. The partitioning as in (4.5) is applied and a modified active line equalizer $\boldsymbol{G}_{\mathrm{aa}}^{(k),\prime}$, which is given by

$$\boldsymbol{G}_{\mathrm{aa}}^{(k),\prime} = \boldsymbol{G}_{\mathrm{aa}}^{(k)} - \boldsymbol{G}_{\mathrm{aj}}^{(k)}\boldsymbol{G}_{\mathrm{jj}}^{(k),-1}\boldsymbol{G}_{\mathrm{ja}}^{(k)}. \tag{4.17}$$

is required to change the number of active receivers.

For the uplink direction, no transmit gain changes are required to apply discontinuous operation. But the bit allocation changes because the noise conditions at the receiver change whenever the equalizer matrix changes.

For zero-forcing equalization $\boldsymbol{G}^{(k)} = \boldsymbol{H}^{(k),-1}$, the receive mean squared error $\mathrm{E}\left[|e_l^{(k),[t]}|^2\right]$ for line l is given by

$$\mathrm{E}\left[|e_l^{(k),[t]}|^2\right] = \sigma^2 g_l^{(k),[t],\mathrm{T}} g_l^{(k),[t],*} \tag{4.18}$$

with noise variance σ^2 and $g_l^{(k),[t],\mathrm{T}}$ to be the lth row of the equalizer matrix $\boldsymbol{G}^{(k),[t]}$. The corresponding configuration-dependent signal-to-noise ratio $SNR_l^{(k),[t]}$ is given by

$$SNR_l^{(k),[t]} = \frac{|s_l^{(k)}|^2}{\sigma^2 g_l^{(k),[t],\mathrm{T}} g_l^{(k),[t],*}}. \tag{4.19}$$

For the more general case that interference is not fully canceled, e.g., due to MMSE equalization as described in Sect. 3.4.2, or due to approximations in the coefficient update, the SNR is given by

$$SNR_l^{(k),[t]} = \frac{|s_l^{(k)}|^2}{\sigma^2 g_l^{(k),[t],\mathrm{T}} g_l^{(k),[t],*} + \sum_{d \neq l} |g_l^{(k),[t],\mathrm{T}} \boldsymbol{h}_d^{(k)} s_d^{(k)}|^2}. \tag{4.20}$$

In both cases, Eqs. (4.19) and (4.20), the gain scaling $\boldsymbol{S}^{(k)}$ is assumed to be constant for different configurations t. The configuration dependent SNR still requires a different bit allocation $\hat{b}_l^{(k),[t]}$ in upstream direction for different configurations.

4.2.6 Bit Allocation and Discontinuous Operation

The signal-to-noise ratio changes for different configurations in uplink as well as in downlink when DO is used. Therefore, the bit allocation is done with respect to the configuration-dependent signal-to-noise ratio $SNR_l^{(k),[t]}$ for line l, carrier k and configuration t.

With a given signal-to-noise-ratio $SNR_l^{(k),[t]}$, the achievable number of bits per channel use is approximated by

$$b_l^{(k),[t]} = \log_2\left(1 + \frac{SNR_l^{(k),[t]}}{\Gamma}\right). \tag{4.21}$$

With the implemented modulation and coding scheme, an integer number of bits $\hat{b}_l^{(k),[t]}$ is modulated on each line l and carrier k during configuration t. The dependency between bit allocation and SNR is described by Eq. (3.16) in Sect. 3.1.5.

For the G.fast case with one NOI bit allocation, $t = 1$, and one DOI bit allocation for $t > 1$, the DOI bit allocation requires further restrictions to achieve the target bit error rate. From the minimum DOI SNR $SNR_{\mathrm{DOI,min},l}^{(k)} = \min_{t:t>1, l \in \mathbb{I}_a^{[t]}} SNR_l^{(k),[t]}$, the bit allocation is approximated by

$$b_{\mathrm{DOI},l}^{(k)} = \log_2\left(1 + \frac{SNR_{\mathrm{DOI,min},l}^{(k)}}{\Gamma}\right). \tag{4.22}$$

while the actual bit allocation $\hat{b}^{(k)}_{\mathrm{DOI},l}$ is derived from the minimum SNR $SNR^{(k)}_{\mathrm{DOI,min},l}$ using the rules from Sect. 3.1.5. The data rate is calculated according to Eq. (4.1) to build the rate matrix $\boldsymbol{R}$.

4.3 Power Minimization by Discontinuous Operation

The main purpose of discontinuous operation is power saving. This section introduces discontinuous operation-based power minimization. Complexity trade-offs and implementation limitations are discussed. This work has been presented in [1, 10].

4.3.1 Scheduling Optimization

The G.fast standard defines a dynamic resource allocation (DRA) entity as part of a DPU [11]. It collects the rate requirements of the individual links, e.g., from buffer fill levels and information about the active services. From that, the required minimum rates $R_{\mathrm{min},l}$ for each line l are derived. They are collected in a minimum rate vector $\boldsymbol{r}_{\mathrm{min}} = [R_{\mathrm{min},1}, \ldots, R_{\mathrm{min},L}]^{\mathrm{T}}$.

According to Sect. 4.1, each TDD frame is partitioned into different configurations $t = 1, \ldots, T$. The objective of framing optimization is firstly, to select a subset of all possible configurations $T \leq 2^L$ which is required to achieve the desired objective, in this case to minimize power consumption.

Secondly, the transmit time $\tau^{[t]}$ for each configuration t is subject to framing optimization. The transmit time $\tau^{[t]}$ defines the fraction of time of the TDD frame, for which the settings associated with configuration t are used, e.g., the transmit gains $\boldsymbol{S}^{(k),[t]}$, the precoder $\boldsymbol{P}^{(k),[t]}$, the bit allocation $\hat{b}^{(k),[t]}_l$ and the set of active lines $\mathbb{I}^{[t]}_{\mathrm{a}}$.

The power consumption $\rho^{[t]}_l$ of line l during configuration t depends on the link state, whether the transmitter is active or discontinued and may also have a dependency on the actual transmit power.

Power consumption minimization subject to rate constraints is formulated as a linear program

$$\min_{\boldsymbol{\tau}} \boldsymbol{c}^{\mathrm{T}}\boldsymbol{\tau} \qquad \text{s.t. } \tau^{[t]} \geq 0 \quad \forall t = 1, \ldots, T \tag{4.23}$$

$$\text{s.t. } \sum_{t=1}^{T} \tau^{[t]} = 1 \text{ s.t. } \boldsymbol{R}\boldsymbol{\tau} \geq \boldsymbol{r}_{\mathrm{min}}.$$

The objective function vector $\boldsymbol{c} \in \mathbb{R}^T$ is

$$c_t = \sum_{l=1}^{L} \rho^{[t]}_l \tag{4.24}$$

which is the sum of the individual power consumption values $\rho^{[t]}_l$ of each line l.

The optimal transmit times for each configuration are mapped to a discrete number of DMT symbols $\hat{\tau}_t$ to be transmitted in each configuration such that the constraints are still satisfied, which is

$$\hat{\tau}^{[t]} = \begin{cases} \left\lceil \tau^{[t]} \cdot N_{\text{sym}} \right\rceil & \text{for } t = 1, \ldots, T-1 \\ 1 - \sum_{t=1}^{T-1} \hat{\tau}^{[t]} & \text{for } t = T \end{cases} \tag{4.25}$$

for N_{sym} DMT symbols in one frame in the respective (upstream or downstream) direction. At that point, a fixed upstream/downstream ratio is assumed and the optimization is performed independently for upstream and downstream direction.

4.3.2 Configuration Selection

In general, the number of possible configurations is $T = 2^L$. Handling the full number of possible configurations for a typical DPU size of 16 or 24 ports would not be manageable in an implementation. This section gives rules how to reduce the required configurations to $T = L + 1$ or $T = L + 2$, which are part of the optimal solution for the scheduling optimization problem.

From a data rate point of view, the sequence in which the configurations are transmitted is not important. Without loss of generality, it can be assumed to start with all lines active, $\mathbb{I}_{\text{a}}^{[1]} = \{1, \ldots, L\}$ as a first configuration and reduce the number of active lines $|\mathbb{I}_{\text{a}}^{[t]}|$ until reaching $\mathbb{I}_{\text{a}}^{[T]} = \{\}$ as the last configuration.

The first proposed scheme, which requires $T = L + 1$ configurations, disables links sequentially with respect to their rate requirements. With the assumption that the rate variation is small $R_l^{[t=1]} \approx R_l^{[t>1]}$, which is ensured by the method described in Sect. 4.3.3, it can be assumed that the transmit time fraction for each link is proportional to $\frac{R_{\min,l}}{R_l^{[t=1]}}$, the ratio between the desired rate and the NOI rate.

The line with the lowest value of $\frac{R_{\min,l}}{R_l^{[t=1]}}$ is discontinued first, while the line with the highest value of $\frac{R_{\min,l}}{R_l^{[t=1]}}$ is discontinued last. This gives a rate matrix $\boldsymbol{R}$ according to

$$\boldsymbol{R} = \begin{bmatrix} R_{\text{NOI},1} & R_{\text{DOI},1} & \ldots & R_{\text{DOI},1} & 0 \\ R_{\text{NOI},2} & R_{\text{DOI},2} & \ldots & 0 & 0 \\ \vdots & & & & \vdots \\ R_{\text{NOI},L} & 0 & \ldots & 0 & 0 \end{bmatrix} \tag{4.26}$$

for the example that line $l = 1$ has the highest value of $\frac{R_{\min,l}}{R_l^{[t=1]}}$ which is reducing to the line $l = 2, \ldots, L$. The example assumes the limitation to two different rates $R_{\text{NOI},l}$ and $R_{\text{DOI},l}$.

Two different DOI-capable precoder architectures, which will give a rate matrix as shown in Eq. (4.26) will be discussed in Sect. 4.3.4.

The discontinuous operation implementation presented in Sect. 4.3.4 uses $T = L + 2$ different configurations and gives a rate matrix according to

$$\boldsymbol{R} = \begin{bmatrix} R_{\text{NOI},1} & R_{\text{DOI},1} & 0 & \dots & 0 \\ R_{\text{NOI},2} & 0 & R_{\text{DOI},2} & \dots & 0 \\ \vdots & & & \ddots & \vdots \\ R_{\text{NOI},L} & \dots & 0 & R_{\text{DOI},L} & 0 \end{bmatrix} \tag{4.27}$$

It is based on the idea that crosstalk is avoided in the DOI by only one active line per configuration t. The number of configurations may be further reduced when low performing configurations are excluded.

4.3.3 Standard-Related Limitations

The G.fast standard requires that only one bit allocation table is used during DOI, which is derived from the minimum SNR for all DOI configurations $t > 1$, $SNR^{(k)}_{\text{DOI},\min,l}$. Therefore, for each line l, there is only the NOI data rate $R_{\text{NOI},l}$ and the DOI data rate $R_{\text{DOI},l}$ which depends on the worst case DOI SNR.

The DOI rate may drop significantly due to a certain selection of active lines $\mathbb{I}_{\text{a}}^{[t]}$, which will then reduce the data rate for all DOI configurations. These "bad configurations" t_{bad} shall be identified in advance and excluded from the set of available configurations. This is the case, e.g., when the DOI precoder as given by Eq. (4.8) is the inverse of an ill-conditioned matrix and requires to scale down transmit power to satisfy the power constraints of Eqs. (4.15) and (4.16). As explained in Sect. 4.2.1, transmission of idle symbols instead of quiet symbols does not require an update of the precoder matrix at the DPU and with that, the critical combination of discontinued lines can be avoided.

In upstream direction, the corresponding receiver remains enabled to avoid a bad configuration.

To identify bad combinations, the individual bit loadings $b_l^{(k),[t]}$ according to Eq. (4.21) are analyzed according to

$$t_{\text{bad}} = \left\{ t : \sum_{k=1}^{K} b_l^{(k),[t]} < \alpha_{\text{bad}} \sum_{k=1}^{K} b_l^{(k),[1]} \right\}. \tag{4.28}$$

which identifies a bad combination such that the data rate drops below $\alpha_{\text{bad}} R_{\text{NOI},l}$. Selecting α_{bad} too strict ($\alpha \approx 1$) reduces the power saving capabilities while selecting it too weak ($\alpha \ll 1$) causes the DOI data rates to drop.

4.3.4 Complexity Limitations

This section derives a scheme to measure the complexity in terms of memory and compute operations associated with the discontinuous operation methods described in previous sections. Compute complexity is measured by the number of compute operations per symbol N_{mac} and memory consumption is measured by the number of data words to be stored N_{mem}.

Without DO, the linear precoder as well as the linear upstream equalizer require to store

$$N_{\text{mem}} = KL^2 \tag{4.29}$$

precoder coefficients.

Tomlinson Harashima precoding requires two matrices to be stored, one full matrix $\boldsymbol{P}_{\text{f}}^{(k)}$ and one lower triangular matrix $\boldsymbol{P}_{\text{b}}^{(k)}$. Therefore, the memory requirement is

$$N_{\text{mem}} = K\left(L^2 + \frac{L(L-1)}{2}\right) \tag{4.30}$$

Linear precoding or linear equalization of one symbol requires

$$N_{\text{mac}} = KL^2 \tag{4.31}$$

multiply-accumulate (MAC) operations to perform the corresponding signal processing steps. For nonlinear precoding, it is

$$N_{\text{mac}} = K\left(L^2 + \frac{L(L-1)}{2} + N_{\text{modulo}}(L)\right) \tag{4.32}$$

where $N_{\text{modulo}}(L)$ represents the compute operations for the modulo operation. These complexity values as summarized in Table 4.1 are compared to the additional complexity required for specific implementations of discontinuous operation.

Table 4.1 Complexity of different crosstalk cancelation methods in terms of compute operations and memory

Mode	Memory	Computation
Linear precoding	KL^2	KL^2
Linear equalization	KL^2	KL^2
TH precoding	$K\left(L^2 + \frac{L(L-1)}{2}\right)$	$K\left(L^2 + \frac{L(L-1)}{2} + N_{\text{modulo}}\right)$

Precoder and Equalizer Coefficient Update (CU)

The first DO implementation discussed updates the precoder, using Eq. (4.8) in every DOI symbol to prepare for the next DOI configuration. This is referred to as coefficient update (CU).

As the precoder and equalizer matrices change from DMT symbol to DMT symbol and there is no time gap between the symbols, there is no computation time between symbols to update the matrices. Therefore, precoder correction requires extra memory to compute the canceler coefficients for the next symbol while the current symbol is processed.

Compared to Eq. (4.29), the memory requirement increases to

$$N_{\mathrm{mem}} = K\left(L^2 + (L-1)^2\right) \tag{4.33}$$

for DOI with coefficient update to store two different linear precoders. For nonlinear precoding, only the forward matrix $\boldsymbol{P}_{\mathrm{f}}^{(k)}$ is doubled, which gives

$$N_{\mathrm{mem}} = K\left(L^2 + (L-1)^2 + \frac{L(L-1)}{2}\right) \tag{4.34}$$

instead of Eq. (4.30). This additional memory gives one symbol time to compute a new precoder coefficients.

Reducing the number of active lines from $L_{\mathrm{a}} + L_{\mathrm{d}}$ lines to L_{a} active lines requires

$$N_{\mathrm{mac}} = L_{\mathrm{a}}^2 L_{\mathrm{d}} + L_{\mathrm{d}}^2 L_{\mathrm{a}} + N_{\mathrm{inv}}(L_{\mathrm{d}}) \tag{4.35}$$

compute operations to evaluate Eq. (4.8) for a precoder update or Eq. (4.17) for the equalizer update. The number of operations for matrix inversion of $\boldsymbol{P}_{\mathrm{dd}}$ or $\boldsymbol{G}_{\mathrm{dd}}$ is $N_{\mathrm{inv}}(L_{\mathrm{d}})$. It depends on the algorithm, but in most cases, matrix inversion has cubic complexity.

For nonlinear precoding, the coefficient update as given by Eq. (4.12) requires

$$N_{\mathrm{mac}} = 2L_{\mathrm{a}}^2 L_{\mathrm{d}} + N_{\mathrm{inv}}(L_{\mathrm{a}}) \tag{4.36}$$

compute operations.

Support of an arbitrary change of the number of active lines L_{a} and discontinued lines L_{d} within one symbol will increase the compute complexity significantly. While the complexity of the precoding operation is of quadratic complexity, $N_{\mathrm{mac}} \sim L^2$, this will turn to $N_{\mathrm{mac}} \sim L^3$ due to the matrix inversion in Eqs. (4.35) and (4.36).

For linear precoding, there exists a special configuration where the coefficient update can be implemented with quadratic complexity, introducing some limitations. This is explained in Appendix C.2. Besides that, using an approximation for matrix inversion gives a discontinuous operation implementation with quadratic complexity.

Discontinuous Operation with Signal Update (SU)

The discontinuous operation implementation according to Sect. 4.3.4 increases the memory and compute requirements for precoding and equalization. This section presents an approach which requires only a small increase in compute and memory complexity, but comes with some performance penalty. A block diagram showing the signal processing required for this method is shown in Appendix C.3.

Precoding in DOI will look like

$$\boldsymbol{y}_{\mathrm{a}}^{(k)} = \boldsymbol{P}_{\mathrm{aa}}^{(k),\prime} \boldsymbol{u}_{\mathrm{a}}^{(k)}. \tag{4.37}$$

This can be rewritten as

$$\boldsymbol{y}_{\mathrm{a}}^{(k)} = \boldsymbol{P}_{\mathrm{aa}}^{(k)} \boldsymbol{u}_{\mathrm{a}}^{(k)} - \boldsymbol{P}_{\mathrm{ad}}^{(k)} \boldsymbol{P}_{\mathrm{dd}}^{(k),-1} \boldsymbol{P}_{\mathrm{da}}^{(k)} \boldsymbol{u}_{\mathrm{a}}^{(k)}. \tag{4.38}$$

With the assumption that the matrix inversion $\boldsymbol{P}_{\mathrm{dd}}^{-1}$ can be approximated by the corresponding first order approximation $\tilde{\boldsymbol{P}}_{\mathrm{dd}}^{(k)} = \boldsymbol{P}_{\mathrm{dd}}^{(k)} - 2\boldsymbol{I}_{L\mathrm{d}} \approx -\boldsymbol{P}_{\mathrm{dd}}^{(k),-1}$, the following simplified approach can be implemented

$$\boldsymbol{y}_{\mathrm{a}}^{(k)} = \boldsymbol{P}_{\mathrm{aa}}^{(k)} \boldsymbol{u}_{\mathrm{a}}^{(k)} + \boldsymbol{P}_{\mathrm{ad}}^{(k)} \tilde{\boldsymbol{P}}_{\mathrm{dd}}^{(k)} \boldsymbol{P}_{\mathrm{da}}^{(k)} \boldsymbol{u}_{\mathrm{a}}^{(k)}. \tag{4.39}$$

With Eq. (4.39), the NOI precoder coefficients can be used during DOI without changes and only the processing steps are changed. The complexity of the modified precoding operation as given by

$$N_{\mathrm{mac}} = (L_{\mathrm{a}}^2 + L_{\mathrm{a}} L_{\mathrm{d}} + L_{\mathrm{d}} L_{\mathrm{d}} + L_{\mathrm{d}} L_{\mathrm{a}})K = L^2 K \tag{4.40}$$

is increased only by a small amount with this method due to the summation of signals in Eq. (4.39). The approximation is based on the assumption that the precoder matrix is close to identity $\boldsymbol{P}_{\mathrm{dd}} \approx \boldsymbol{I}$. To implement a similar approximation for uplink equalization, the equalizer $\boldsymbol{G}^{(k)}$ shall also be normalized to unit diagonal. This can be done by dividing the equalizer into two parts, a diagonal equalizer $\boldsymbol{G}_{\mathrm{diag}}^{(k)}$ and a scaled equalizer matrix $\boldsymbol{G}_{\mathrm{xt}}^{(k)}$. They are given by

$$\boldsymbol{G}_{\mathrm{diag}}^{(k)} = \mathrm{diag}\left(\mathrm{diag}\left(\boldsymbol{G}^{(k)}\right)\right) \tag{4.41}$$

$$\boldsymbol{G}_{\mathrm{xt}}^{(k)} = \boldsymbol{G}_{\mathrm{diag}}^{(k),-1} \boldsymbol{G}^{(k)}. \tag{4.42}$$

This gives the following signal flow in the uplink equalizer

$$\hat{\boldsymbol{u}}_{\mathrm{a}}^{(k)} = \boldsymbol{G}_{\mathrm{diag,aa}}^{(k)} \left(\boldsymbol{G}_{\mathrm{xt,aa}}^{(k)} \hat{\boldsymbol{y}}_{\mathrm{a}}^{(k)} + \boldsymbol{G}_{\mathrm{xt,ad}}^{(k)} \tilde{\boldsymbol{G}}_{\mathrm{xt,dd}}^{(k)} \boldsymbol{G}_{\mathrm{xt,da}}^{(k)} \hat{\boldsymbol{y}}_{\mathrm{a}}^{(k)} \right). \tag{4.43}$$

where $\tilde{\boldsymbol{G}}_{\mathrm{xt,dd}}^{(k)} = \boldsymbol{G}_{\mathrm{xt,dd}}^{(k)} - 2\boldsymbol{I}_{L\mathrm{d}}$ and $\hat{\boldsymbol{y}}_{\mathrm{a}}^{(k)}$ is the receive signal of the active lines before equalization. No additional memory is required for the modified precoding operation.

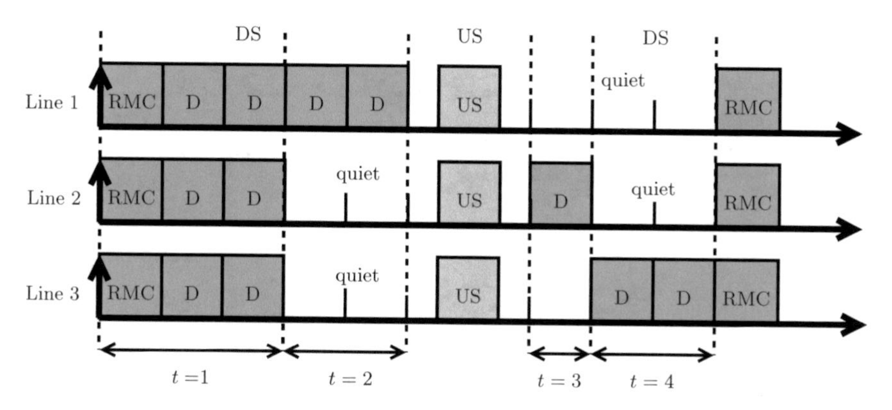

Fig. 4.5 Crosstalk avoidance in discontinuous operation

Table 4.2 Additional complexity due to discontinuous operation

Mode	Memory	Computation
LP coefficient update (CU)	$(L-1)^2 K$	$L(L-1)K$
LP signal update (SU)	–	–
Crosstalk avoidance (CA)	–	–
THP coefficient update	$(L-1)^2 K$	$2(L-1)^2 + N_{\text{inv}}(L-1)$

Crosstalk Avoidance (CA) in DO

A special mode of discontinuous operation completely avoids crosstalk within the DOI by separating individual subscribers over time, as shown in Fig. 4.5. DO crosstalk avoidance combines time division multiple access (TDMA) with crosstalk cancelation by switching between both in time. Due to the fact that precoding is not used in the DOI, this method is applicable to linear and non-linear precoding.

The rate matrix for crosstalk avoidance is given by Eq. (4.27). With this scheme, no precoding is required in the DOI phase while individual links still achieve high peak data rates in case that the other lines do not require high data traffic. Table 4.2 summarizes the additional complexity required for different DO implementation methods.

4.4 Peak Rate Optimization with Discontinuous Operation

Even though discontinuous operation has been introduced to save power, it is possible to increase data rates of individual users temporarily by using discontinuous operation. The idea of giving one channel exclusively to a single subscriber is presented in Sect. 4.3.4 for power saving purposes.

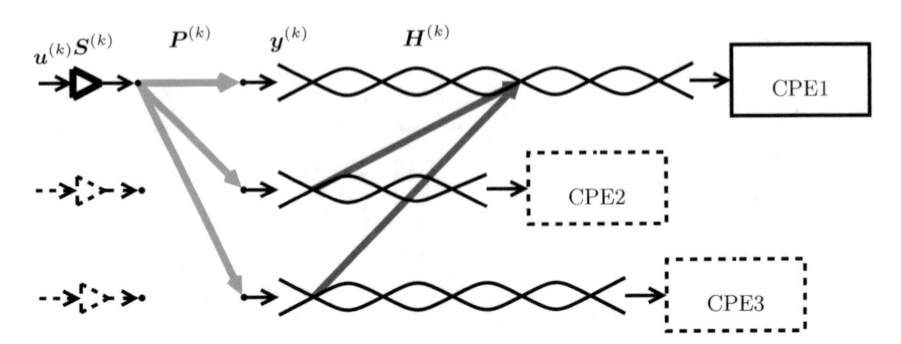

Fig. 4.6 Only CPE 1 is served at a certain time frame, using the transmitters of all lines

Instead of shutting down the transmitters of the inactive lines in the DOI state, these transmitters may be used to improve the data rate of the active lines [12], e.g., in the configuration shown in Fig. 4.6.

Figure 4.6 shows the case where only a single line is served. By that, the highest peak rates are available to individual subscribers. Other operating points, where other subsets of lines are served in the DOI are also possible and require non-square precoder matrices.

4.4.1 Beamforming-Based Precoding

For the case of Fig. 4.6, zero-forcing precoding based on the Moore-Penrose pseudoinverse according to Eq. (3.48), is not applicable, as there is no crosstalk to be canceled. For an optimized usage of the precoder, weighted MMSE Precoding, as described in Sect. 3.2.9 or Sect. 3.2.10 can be applied. Besides the weighted MMSE approach or a precoder calculation based on generalized inverses, as shown in Appendix C.4, the precoder coefficients can be derived in a way which is similar to beamforming in the wireless case.

The precoder coefficients for achieving the peak rate on line v are

$$p_{dv}^{(k)} = \mathrm{e}^{-j\angle h_{vd}^{(k)}} \tag{4.44}$$

for $d = 1, \ldots, L$ from the channel matrix coefficient $h_{vd}^{(k)} = |h_{vd}^{(k)}|\mathrm{e}^{j\angle h_{vd}^{(k)}}$.

Even though, there are $T = L + 2$ different configurations required to offer the peak rate to all L subscribers, each DOI precoder consists of a single column and therefore, the additional memory required is

$$N_{\text{mem}} = KL^2, \tag{4.45}$$

which doubles the memory requirements compared to precoding without peak rate optimization. The compute requirements in the DOI are lower than in the NOI.

4.4.2 *Spectrum Optimization*

Spectrum optimization is required for precoder coefficients derived according to Eq. (4.44). As there is only one line active at a time, the optimization is very similar to the single line spectrum optimization case discussed in Sect. 3.1.6. The optimization problem changes from Eq. (3.23) to

$$\min_{x_l^{(k)}, k=1,\dots,K} \sum_{k=1}^{K} \log_2 \left(1 + \frac{\left| \boldsymbol{h}_l^{(k),\mathrm{T}} \boldsymbol{p}_l^{(k),[t]} \right|^2 x^{(k)}}{\sigma^{(k),2}\Gamma} \right) \tag{4.46}$$

$$\text{s.t.} \sum_{k=1}^{K} x_l^{(k)} \leq p_{\text{sum}}$$
$$\text{s.t. } x_l^{(k)} \leq p_{\text{mask}}^{(k)}$$
$$\text{s.t. } x_l^{(k)} \leq p_{\text{bmax}}^{(k)}$$
$$\text{s.t. } x_l^{(k)} \geq 0$$

for each line l independently where $\boldsymbol{p}_l^{(k),[t]}$ is the precoder column to serve line l in configuration t. Besides that, the calculation of $p_{\text{bmax}}^{(k)}$ changes from Eq. (3.27) to

$$p_{\text{bmax}}^{(k)} = \left(2^{b_{\max}} - 1\right) \frac{\sigma^{(k),2}\Gamma}{\left| \boldsymbol{h}_l^{(k),\mathrm{T}} \boldsymbol{p}_l^{(k),[t]} \right|^2}. \tag{4.47}$$

The single gain scaling coefficient $s_l^{(k),[t]}$ for DOI, $t > 1$, is given by $s_l^{(k),[t]} = \sqrt{x_l^{(k)}}$ where only line l is served during configuration t. In case that the sum-power constraint is active, this method is suboptimal, compared to the MMSE solution, but the computational complexity is much lower.

4.4.3 *Scheduling Optimization for Peak Rate*

Similar to the crosstalk avoidance case, the peak rate optimization leads to a rate matrix $\boldsymbol{R}$ according to Eq. (4.27), while the objective changes from power consumption minimization to peak rate maximization.

Scheduling optimization for peak rates requires to change the optimization problem to be solved with respect to the lines requesting a high peak rate. Therefore, the set $\mathbb{I}_{\text{peak}}$ is introduced, containing the lines which request the peak rate. Besides that, a minimum rate vector $\boldsymbol{r}_{\min}$ as defined in Sect. 4.3.1 contains the minimum rate for all lines, including the lines in $\mathbb{I}_{\text{peak}}$.

Three cases are distinguished.

1. There are no lines requesting the peak rate, $\mathbb{I}_{\text{peak}} = \{\}$.
2. One line request a high peak rate, $|\mathbb{I}_{\text{peak}}| = 1$.
3. Multiple lines request a high peak rate, $|\mathbb{I}_{\text{peak}}| > 1$.

Scheduling optimization is a linear program of the form

$$\min_{\boldsymbol{\tau}} \boldsymbol{c}^{\mathrm{T}}\boldsymbol{\tau} \qquad \text{s.t. } \tau^{[t]} \geq 0 \quad \forall t = 1, \ldots, T \qquad (4.48)$$
$$\text{s.t. } \sum_{t=1}^{T} \tau^{[t]} = 1 \qquad \text{s.t. } \boldsymbol{R\tau} \geq \boldsymbol{r}_{\min}$$

for all three cases, while the objective function vector changes. For case 2, the typical case of the peak rate optimization, it is given by

$$c_l = -R_{vl} \tag{4.49}$$

to maximize the peak rate on a single line $\mathbb{I}_{\text{peak}} = \{v\}$.

For the third case that multiple lines request the peak rate, it is

$$c_t = \sum_{v \in \mathbb{I}_{\text{peak}}} -R_{vt} \tag{4.50}$$

which usually leads to the use of the NOI configuration throughout the whole TDD frame.

For the first case that no line requests a high peak rate, the objective may be switched to power minimization according to Eq. (4.24) in Sect. 4.3.1.

4.5 Discussion

Framing-based rate optimization gives an additional degree of freedom for physical layer optimization in copper line systems besides precoding and spectrum optimization, discussed in Chap. 3. The presented framing optimization methods aim to solve a linear program to achieve certain target rates or minimize power consumption. The linear program can be solved efficiently in short time which allows to react quickly on changing requirements, e.g., changes of the requested data rates of individual subscribers.

Discontinuous operation, the ability to shut down transmitter and receiver components on a DMT-symbol basis, has been introduced in G.fast for power saving, which is the primary application of framing optimization discussed.

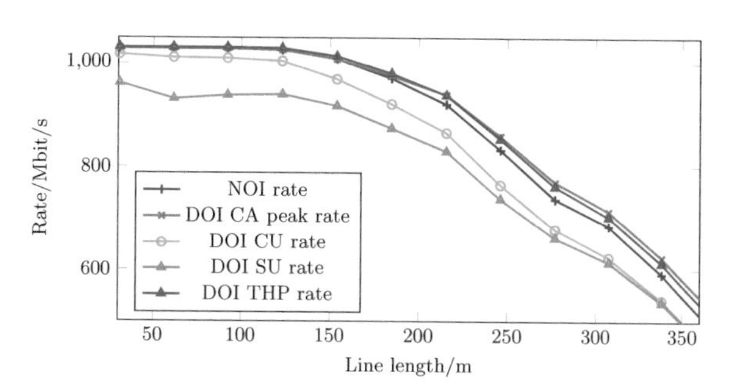

Fig. 4.7 Rate versus reach for different discontinuous operation implementations for 16 active lines on a DPU

4.5.1 Power Minimization

The motivation for discontinuous operation as the main tool to save power in G.fast is given in Appendix C.5.

At this point, the power saving capabilities are not measured in terms of power, but in terms of the ratio between the time when the transmitter is active and the over-all transmit time. This allows an implementation-independent evaluation of the power saving capabilities of discontinuous operation by choosing $\rho_l^{[t]} = 0$ for quiet symbols and $\rho_l^{[t]} = 1$ for idle and data symbols. The per-line average power consumption is $\rho_l = \sum_{t=1}^{T} \tau_t \rho_l^{[t]}$. $\rho_l = 1$ corresponds to a link that is always active.

Even though the idea of discontinuous operation is that the power consumption scales linearly with the desired data rates, this target is not achieved exactly with discontinuous operation implementations for several reasons. Mainly, because

- the achieved data rates in DOI are lower than in NOI and
- a line cannot be discontinued at any time due to constraints in the available configurations or constraints on the switching time.

At this point, systems with $L = 16$ lines are investigated. In Appendix C.6, simulation results for different systems sizes are shown. Figure 4.7 shows the DOI data rates for different DO implementations in comparison with the NOI data rate using linear precoding. The highest data rates are achieved by DO crosstalk avoidance (CA) (see Sect. 4.3.4), as only one link is active at a time.[4] However, these rate is only available to one subscriber at a time. THP-based DO also achieves high rates because the THP data rates are higher, in general. Discontinuous operation implementations based on coefficient update (CU), as described in Sect. 4.3.4, achieve lower rates due to the standardization limitation to one bits and gains table in the DOI and the lowest rates are achieved by signal update (SU) due to the approximated matrix inversion (see Sect. 4.3.4).

[4]In contrast to peak rate optimization, as discussed in Sect. 4.4 where multiple transmitters are active to serve one subscriber (MISO), DOI CA uses a single transmitter to serve a single receiver (SISO), while the other transmitters are disabled to save power.

Fig. 4.8 Average link on-time over average requested rates for a DPU with 16 lines

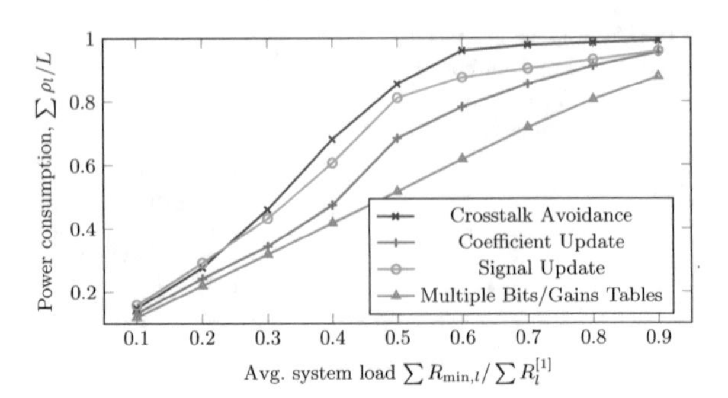

Higher DOI data rates allow to transport the required data at a shorter time, increasing the power saving capabilities. The power saving capabilities in terms of the average link on-time is shown in Fig. 4.8 as a function of the average load on the DPU. The average load is determined by the ratio between the sum of desired rates over the sum of NOI rates $\frac{\sum R_{\mathrm{min},l}}{\sum R_l^{[1]}}$ while the individual rates $R_{\mathrm{min},l}$ are different for each line l [1].

Figure 4.7 shows that a discontinuous operation implementation without the G.fast-related limitation to one bits and gains table in DOI gives the highest power saving and the average power consumption of the DPU scales linearly with the average requested data rates. When using the coefficient update method described in Sect. 4.3.4, the data rates in the DOI are lower than in the NOI. Therefore, more time is required to transmit the data, which reduces the power saving.

The signal update scheme described in Sect. 4.3.4 achieves even lower DOI data rates, requiring more transmit time. The crosstalk avoidance scheme, which is the lowest complexity scheme, gives the lowest power saving at high loads, but at the very low loads of 20 % and below, it shows advantages due to the high peak rates achieved.

4.5.2 Peak Rates

Due to the high crosstalk on the G.fast channels, there is a significant gap between single line rates and line rates in the MIMO case at the sum-rate optimal point, as explained in Appendix B.15. Observing the data traffic of multiple lines over time, the sum-rate optimal point is not always the optimal point from a user perspective, as not all lines require the maximum rate at the same time. With knowledge of the current data traffic requirements and the capability to adjust the precoder settings to the current requirement at sufficient speed, it is possible to serve subscribers at much higher rates than they are achievable at the sum-rate optimal point.

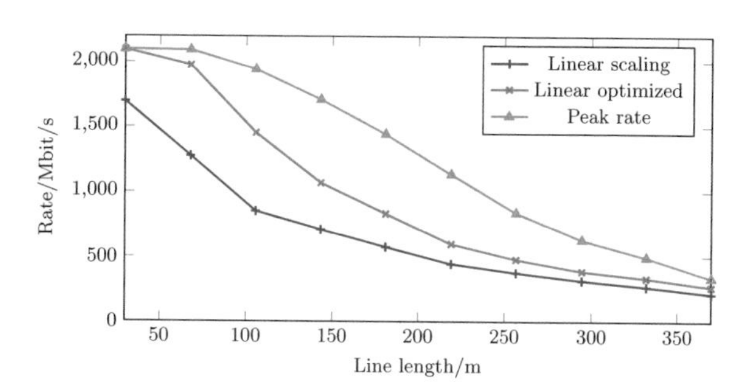

Fig. 4.9 Rate versus reach curves for comparison between achievable rates for sum-rate optimization and achievable peak rates on a 30 pair DTAG PE06 binder with non-co-located CPEs

Peak rate optimization allows to use this degree of freedom with the result that the achievable peak rates of individual lines depend on the data traffic on the other lines.

Figure 4.9 shows the achievable peak rates in comparison to the rates at the sum rate optimal precoding achieved with linear column norm scaling (Sect. 3.2.6) and spectrum optimized linear zero-forcing precoding (Sect. 3.2.8). Appendix C.7 gives more details on the simulation conditions.

The peak rates in Figure 4.9 are achieved with zero base load on the other lines, e.g., their target rates are zero. With a certain base load on the other lines, the peak rate gain reduces accordingly and comes closer to the sum-rate optimal case.

The observed performance gains are significant, especially for the medium and longer loops. Appendix C.7 shows more simulation results and gives a comparison between peak rate and single line rate on the different network topologies.

References

1. Strobel, R., Utschick, W.: Discontinuous operation for precoded G.fast. In: International Conference on Acoustics, Speech and Signal Processing (ICASSP). IEEE (2016)
2. Oksman, V., Strobel, R., Bry, C.: Dynamic update of transmission settings and robust management communication channel (2013). EU Patent App. EP2661009
3. Oksman, V., Strobel, R.: G.fast: additional functionality associated with discontinuous operation (2013). ITU-T Contribution SG15/Q4a t 2013-05-Q4-058
4. Strobel, R., Oksman, V.: Methods, devices and systems of supporting discontinuous operation in communication systems using vectoring (2016). US Patent App. 14/890,173
5. Strobel, R., Smaoui, L., Oksman, V.: Low power modes for data transmission from a distribution point (2016). US Patent App. 20/160,087,677
6. Oksman, V., Strobel, R.: G.fast: precoder update in support of discontinuous operation (2013). ITU-T Contribution SG15/Q4a 2013-01-Q4-068
7. Oksman, V., Strobel, R.: G.fast: issues with discontinuous operation (2013). ITU-T Contribution SG15/Q4a t 2013-05-Q4-057
8. Maes, J., Nuzman, C.: Energy efficient discontinuous operation in vectored G.fast. In: IEEE International Conference on Communications (ICC), pp. 3854–3858. IEEE (2014)
9. ITU-T Rec. G.9700: Fast access to subscriber terminals (FAST) - power spectral density specification. ITU Recommendation (2013)

10. Strobel, R.: Discontinuous operation for G.fast in the DP (2016). G.fast Summit 2016. http://www.uppersideconferences.com/g.fast/-gfast_2016_agenda_conference_day_2.html
11. ITU-T Rec. G.9701: Fast access to subscriber terminals - physical layer specification. ITU Recommendation (2015)
12. Huang, Y., Magesacher, T., Medeiros, E., Lu, C., Eriksson, P.E., Odling, P.: Rate-boosting using strong crosstalk in next generation wireline systems. In: IEEE Global Communications Conference (GLOBECOM), pp. 1–6. IEEE (2015)

Chapter 5
Conclusion

This thesis is based on the work for development of G.fast and the fiber to the distribution point network. The new technology is derived from VDSL2 and Vectored VDSL2 and re-uses many of the techniques introduced there. Some aspects in the G.fast physical layer experienced fundamental changes, compared to previous wireline access network technologies.

The G.fast Technology

The massive increase of bandwidth to 106 MHz and even 212 MHz requires a different approach to crosstalk management than just "canceling" the crosstalk, as it is done in Vectored VDSL2. Besides that, changing the duplexing scheme to TDD and further introducing discontinuous operation in a multi-user MIMO environment is another main innovation of G.fast.

While the increased bandwidth serves the demand for higher and higher data rates, discontinuous operation is motivated by power saving, as a low power consumption is a key to make FTTdp networks possible and bring the fiber closer to the subscribers, where no local power supply is available at the DP.

The field of optimized precoding and spectrum optimization gives some interesting scientific problems to be solved. A major goal of the presented work is that scientific results are created with implementation aspects in mind. With this objective, it turns out that some known results from research do not turn into reality under specific implementation conditions such as the transmission channels, the power and spectral constraints and the complexity limitations that come into play under real-world conditions.

Channel Models

With this goal, a good understanding of the G.fast transmission channel as it is presented in Chap. 2 is an important basis of the following work on physical layer optimization. A deep understanding of the channel characteristics and the ability to

R. Strobel, *Channel Modeling and Physical Layer Optimization in Copper Line Networks*, Signals and Communication Technology,
https://doi.org/10.1007/978-3-319-91560-9_5

create realistic high frequency channels for simulation helps to support technology decisions in the standard and the development of the technology.

The presented models are part of the industry standard TR-285 [1] which defines channel models for the fiber to the distribution point network. The models supported many technology decisions on G.fast DMT modulation parameters [2], channel estimation [3–6], precoding and equalization methods [7] and the link training and adaptation methods [8].

Precoding and Spectrum Optimization

With more knowledge on the G.fast channels and observations which are not present for VDSL2 frequencies, e.g., that the crosstalk channel on a copper cable binder may have less attenuation than the direct copper connection, precoding and spectrum optimization methods require major changes compared to the VDSL2 Vectoring system.

In Chap. 3, the precoding and spectrum optimization problem is investigated from different angles. The approach of optimized zero-forcing precoding, as presented in Sects. 3.2.8 and 3.2.10, follows the idea of continuous improvement of the existing VDSL2 Vectoring approach with knowledge about the specific characteristics of the G.fast channel and the G.fast system. Tomlinson–Harashima precoding, which has been one of the technology candidates for VDSL2 Vectoring [9], is another example of this improvement approach, discussed in Sect. 3.3.3.

MMSE precoding methods [10, 11] result from another approach of using results from the research on MIMO precoding in wireless communication. MMSE precoding for G.fast is investigated in Sects. 3.2.9 and 3.2.10. While the approach presented in Sect. 3.2.9 is obviously not practical due to the computational complexity of the coefficient calculation, the gradient method from Sect. 3.2.10 shows advantages for the tracking phase.

These methods, which represent implementable precoding and spectrum allocation schemes and which are investigated with respect to the G.fast implementation limitations, are compared with channel capacity and theoretical upper bounds as derived in Sect. 3.5.

The application of convex optimization methods on rate optimization for G.fast, which is in general a discrete and thus a combinatorial problem, is proven to be a very effective tool to derive efficient algorithms. The comparison with channel capacity, which is again a continuous objective, shows optimized linear and non-linear zero-forcing precoding performance is close to capacity.

Discontinuous Operation

With discontinuous operation, G.fast introduces the ability to use time-dependent optimization and switch quickly between different system settings at a time scale where the subscriber cannot notice it. This allows to operate the links virtually even beyond the channel capacity limits, as each individual subscriber can be given a much higher peak rate whenever it is requested on the link, as described in Sect. 4.4.

Besides that, the main purpose of discontinuous operation is to save power, and scale the power consumption with the current data traffic, again, invisible for the subscriber, because the data rate is available when it is requested.

Outlook

Even though twisted pair-based access networks are considered a technology that is replaced by fiber to the home networks [12], the difficulties observed in fiber deployments and the cost disadvantage of fiber-only networks compared to hybrid fiber-copper networks [13] show that the development of copper-based access is not at the end with G.fast.

After the step from 106 to 212 MHz G.fast, further performance and latency improvement steps such as full duplex operation and bandwidth extension towards 500 MHz show that copper-based technologies will also be able to compete with fiber in future. The convergence of coax-based access networks and twisted pair-based access networks is another trend observed in future copper access networks, as it is seen in developments such as G.fast over coax [14] or XG.fast and XG.cable [15].

The demand for higher throughput, higher data rates and lower latency in the access network is still not at the end and will continue to drive innovation and research.

References

1. Broadband Forum TR-285: Cable Models for Physical Layer Testing of G.fast Access Network. Technical report (2015)
2. Oksman, V., Strobel, R.: G.fast: update for the guard interval value (2012). ITU-T Contribution SG15/Q4a 2012-06-4A-054
3. Oksman, V., Verbin, R., Strobel, R., Keren, R.: G.fast: Channel estimation method for G.fast (2013). ITU-T Contribution SG15/Q4a 2013-01-Q4-039
4. Oksman, V., Strobel, R.: G.fast: performance of channel estimation methods up to 200 MHz (2012). ITU-T Contribution SG15/Q4a t 2012-11-4A-054
5. Oksman, V., Strobel, R.: G.fast: impact of interpolation on convergence process during initialization (2013). ITU-T Contribution SG15/Q4a 2013-09-Q4-064
6. Oksman, V., Strobel, R.: G.fast: performance of channel estimation methods (2013). ITU-T Contribution SG15/Q4a COM 15 - C 2236 -E
7. Oksman, V., Strobel, R.: G.fast: performance evaluation of linear and non-linear precoders (2013). ITU-T Contribution SG15/Q4a 2013-01-Q4-041R1
8. Oksman, V., Strobel, R.: G.fast: proposal to define a synchronous multi-line OLR (2013). ITU-T Contribution SG15/Q4a 2013-05-Q4-063
9. Ginis, G., Cioffi, J.: Vectored transmission for digital subscriber line systems. IEEE J. Sel. Areas Commun. **20**(5), 1085–1104 (2002)
10. Bogale, T.E., Vandendorpe, L.: Sum MSE optimization for downlink multiuser MIMO systems with per antenna power constraint: Downlink-uplink duality approach. In: International Symposium on Personal Indoor and Mobile Radio Communications (PIMRC), pp. 2035–2039. IEEE (2011)
11. Joham, M., Utschick, W., Nossek, J., et al.: Linear transmit processing in MIMO communications systems. IEEE Trans. Signal Process. **53**(8), 2700–2712 (2005)

12. Weldon, M.K., Zane, F.: The economics of fiber to the home revisited. Bell Labs Tech. J. **8**(1), 181–206 (2003)
13. Timmers, M., Guenach, M., Nuzman, C., Maes, J.: G.fast: evolving the copper access network. IEEE Commun. Mag. **51**(8) (2013)
14. Bittancourt, W.: Benefits and Options to Deploy G.fast over COAX (2016). G.fast Summit 2016, http://www.uppersideconferences.com/g.fast/-gfast_2016_agenda_conference_day_1.html
15. Coomans, W., Moraes, R.B., Hooghe, K., Duque, A., Galaro, J., Timmers, M., van Wijngaarden, A.J., Guenach, M., Maes, J.: XG-fast: the 5th generation broadband. IEEE Commun. Mag. **53**(12), 83–88 (2015)

Appendix A

A.1 Parameters for Single Line Models

This section summarizes single line model parameters for G.fast cable types. Some of the cables are part of the ITU G.fast standard [1] or the FTTdp cable modeling standard [2]. The single line models are described in Sect. 2.1.1.

The ITU single line model [1] is characterized by

$$Z_s(j\omega) = j\omega L_{s\infty} + R_{s0}\left(1 - q_s q_x + \sqrt{q_s^2 q_x^2 + 2\frac{j\omega}{\omega_s}\left(\frac{q_s^2 + j\omega/\omega_s q_y}{q_s^2/q_x + j\omega/\omega_s q_y}\right)}\right), \tag{A.1}$$

and

$$Y_p(j\omega) = j\omega C_{p0}(1 - q_c)\left(1 + \frac{j\omega}{\omega_d}\right)^{-2\phi/\pi} + j\omega C_{p0} q_c \tag{A.2}$$

with the additional definitions

$$L_{s\infty} = \frac{1}{\eta_{VF} c_0} Z_{0\infty}\,,\ C_{p0} = \frac{1}{\eta_{VF} c_0}\frac{1}{Z_{0\infty}}, \tag{A.3}$$

$$q_s = \frac{1}{q_H^2 q_L}, \omega_s = q_H^2\left(\frac{4\pi R_{s0}}{\mu_0}\right). \tag{A.4}$$

The parameters for some cable types used for G.fast evaluation are summarized in Table A.1. The ITU-CAD55 cable is defined in the G.fast standard [1] and reused in [2]. It refers to a British Telecom [3] cable. The cable type named DTAG-PE05 as well as the DTAG-PE06 cable are based on measurements at Deutsche Telekom. The DTAG-PE05 cable is called I-O2YS(ST)H 10x2x0.5 STVI Bd in the standard [2] while the DTAG-PE06 cable is named J-H(ST)H Bd 10x2x0.6.

R. Strobel, *Channel Modeling and Physical Layer Optimization in Copper Line Networks*, Signals and Communication Technology,
https://doi.org/10.1007/978-3-319-91560-9

Table A.1 Parameters of the ITU model for different cable types from the ITU G.9701 [1] and the Broadband Forum TR-285 [2] standard

Parameter	ITU-CAD55	DTAG-PE05	DTAG-PE06
R_{s0}	0.1871	0.2145	0.0867
η_{VF}	0.6976	0.7489	0.6812
$Z_{0\infty}$	105.0694	134.2196	110.8
q_H	0.7415	0.7415	0.50105
q_L	1.5315	2.160729447	4.0
q_x	1	1	1
q_y	0	0	0
ϕ	−0.2356	−0.0319	0.0164
q_c	1.0016	1.0016	1.0016
f_d	1	1	1

Table A.2 Parameters of the ITU model for additional cable types, which are not part of [1] or [2]

Parameter	DTAG-PE04	DTAG-YSTY	HSA-YSTY
R_{s0}	0.3	0.1346	0.6201
η_{VF}	0.6176	0.6079	0.7136
$Z_{0\infty}$	129.85	70.9305	81.0305
q_H	0.7687	0.60105	0.7965
q_L	1.8630	9.17386	2.1446
q_x	1	1	1
q_y	0	0	0
ϕ	0.00286	−0.1241	−0.19411
q_c	0.00259	1.0202	1.0016
f_d	1	1	1

Table A.2 summarizes parameters for some additional cable types, which are not part of the cable modeling standards, but have been measured as part of the measurement campaigns [4–6]. The cable named DTAG-PE04 is a lower quality access network cable of Deutsche Telekom and the DTAG-YSTY as well as the HSA-YSTY are in-house cables.

The ETSI BT0 model is an alternative single line model for higher frequencies. It is characterized by the equations

$$Z_s(f) = \sqrt[4]{R_{0c}^4 + a_c f^2} + j\omega \left(\frac{L_0 + L_\infty (f/f_m)^{N_b}}{1 + (f/f_m)^{N_b}} \right) \tag{A.5}$$

Table A.3 Parameters of the ETSI BT0 model [7] for different cable types as defined in the Broadband Forum TR-285 standard [2]

	DTAG-PE05	DTAG-PE06
R_{0c}	20.1699	160.16183
a_c	0.0429	0.03329
L_0	0.00090878	0.00090679
L_∞	0.0005999	0.0005342
N_b	0.7139	0.7218
f_m	91895.48	91895.48
g_0	4.342e-12	23.7312e-12
N_{ge}	1.10544	1.1518
C_0	7.7605e-08	7.7607e-08
C_∞	3.2378e-08	4.6143e-08
N_{ce}	0.5674	0.5463

and

$$Y_p(f) = \left(g_0 f^{N_{ge}}\right) + j\omega \left(C_\infty + \frac{C_0}{f^{N_{ce}}}\right). \tag{A.6}$$

Parameters for the DTAG-PE05 and DTAG-PE06 cable can be found in Table A.3. They are part of the TR-285 cable modeling standard [2].

A.2 Parameters for Crosstalk Models

The ATIS model, defined in [8], as well as the TR-285 Annex A crosstalk model in [2] are based on a statistical model of the average crosstalk attenuation. The models are described in Sect. 2.1.2.

The crosstalk coupling path from disturber line d to victim line v is described as

$$H_{\text{FEXT}\,vd}(f) = |H_v(f)|\, f e^{j\varphi(f)} \kappa \sqrt{z_{\text{coupling}}}\, 10^{x_{\text{dB}\,vd}/20} \tag{A.7}$$

where the crosstalk statistics are part of the random matrix $\boldsymbol{X}_{\text{dB}}$. The logarithmic values of the matrix follow a beta distribution with the probability density function [8]

$$f(x) = \frac{(x-a)^{\alpha-1}(b-x)^{\beta-1}}{\mathrm{B}(\alpha,\beta)(b-a)^{\alpha+\beta-1}} \tag{A.8}$$

where $\mathrm{B}(\alpha, \beta)$ is the beta function.

The parameters are summarized in Table A.4. In TR-285 Annex A, a 24×24 reference matrix is defined instead of a statistical model.

Table A.4 Parameters for the ATIS NIPP-NAI [8] and the TR-285 Annex A [2] model

Parameter	ATIS
a	-60 dB
b	10 dB
α	11 dB
β	6.6 dB
κ	$1.594 \cdot 10^{-10}$

A.3 Parameters for Spatial Domain Model

The spatial domain MIMO model from Sect. 2.2 requires some geometry parameters for crosstalk modeling, which are summarized in this section. The wires are organized as twisted pairs or star quads with four wires, which are then grouped to a sub-binder. Multiple sub-binders are grouped to the cable binder, depending on the size of the binder. The DTAG-PE04, DTAG-PE05 and DTAG-PE06 cable are quad cables where 5 quads form a sub-binder and the overall number of pairs in the binder is a multiple of $L_{\text{sub}} = 10$. The DTAG-PE05 and the DTAG-PE04 cable are star quad cables with continuous twist. The DTAG-PE06 cable is a star quad cable with alternating twist, i.e, several meters of the cable are twisted in one direction and after that, the twist direction is inverted (Tables A.5 and A.6).

The YSTY cables as well as the BT CAD55 cable type are twisted pair cables. While the YSTY cables do not have a sub-binder structure, the CAD55 cable is organized in 10-pair sub-binders, $L_{\text{sub}} = 10$.

Table A.5 Parameters for cable geometry of TR-285 Annex B model [2]

Parameter	DTAG-PE04	DTAG-PE05	DTAG-PE06
r_{ik}	1.1314 mm	1.35 mm	1.1532 mm
r_{binder}	2.7 mm	3 mm	2.65 mm
r_i	0.2 mm	0.25 mm	0.3 mm
d_{twist}	0.0476 m	0.0476 m	0.055 m
	0.05 m	0.05 m	0.07 m
	0.0525 m	0.0525 m	0.08 m
	0.0564 m	0.0564 m	0.09 m
	0.063 m	0.061 m	0.105 m
σ_{lrd}	$1.8 \cdot 10^{-8}$	$2 \cdot 10^{-8}$	$2.1 \cdot 10^{-8}$
σ_{lrx}	$1.4 \cdot 10^{-9}$	$1 \cdot 10^{-9}$	$1 \cdot 10^{-9}$

Table A.6 More parameters for cable geometry of the spatial domain MIMO model

Parameter	DTAG YSTY	HSA YSTY	BT CAD55
r_{ik}	1.1532 mm	1.1532 mm	1.5 mm
r_{binder}	2.6 mm	2.9 mm	2.8 mm
r_i	0.3 mm	0.3 mm	0.25 mm
d_{twist}	0.11 m	0.1 m	0.1 m
σ_{lrd}	$3 \cdot 10^{-8}$	$4 \cdot 10^{-9}$	$3 \cdot 10^{-8}$
σ_{lrx}	$3 \cdot 10^{-9}$	$3 \cdot 10^{-9}$	$6 \cdot 10^{-9}$

For the DTAG-PE04, DTAG-PE05 and DTAG-PE06 cable, the wire position $x_{\text{wire},i}(z)$ in space as a function of line length z is given by

$$\begin{bmatrix} x_{\text{wire},1}(z) \\ x_{\text{wire},2}(z) \\ x_{\text{wire},3}(z) \\ x_{\text{wire},4}(z) \end{bmatrix} = h_x + r_{\text{binder}} \cos\left(2\pi \left\lceil \frac{i}{4} \right\rceil\right) + r_{ik} \begin{bmatrix} \sin\left(2\pi \frac{z}{d_{\text{twist},\lceil i/4 \rceil}}\right) \\ -\sin\left(2\pi \frac{z}{d_{\text{twist},\lceil i/4 \rceil}}\right) \\ \cos\left(2\pi \frac{z}{d_{\text{twist},\lceil i/4 \rceil}}\right) \\ -\cos\left(2\pi \frac{z}{d_{\text{twist},\lceil i/4 \rceil}}\right) \end{bmatrix} \tag{A.9}$$

and $y_{\text{wire},i}(z)$ by

$$\begin{bmatrix} y_{\text{wire},1}(z) \\ y_{\text{wire},2}(z) \\ y_{\text{wire},3}(z) \\ y_{\text{wire},4}(z) \end{bmatrix} = h_c + r_{\text{binder}} \sin\left(\frac{2\pi}{5} \left\lceil \frac{i}{4} \right\rceil\right) + r_{ik} \begin{bmatrix} \cos\left(2\pi \frac{z}{d_{\text{twist},\lceil i/4 \rceil}}\right) \\ -\cos\left(2\pi \frac{z}{d_{\text{twist},\lceil i/4 \rceil}}\right) \\ -\sin\left(2\pi \frac{z}{d_{\text{twist},\lceil i/4 \rceil}}\right) \\ \sin\left(2\pi \frac{z}{d_{\text{twist},\lceil i/4 \rceil}}\right) \end{bmatrix}. \tag{A.10}$$

where the constants h_x and h_c are used to place the different sub-binders in space.

A.4 Summary on Cable Measurement Data for MIMO Modeling

This section shows FEXT and direct channel transfer functions of cables measured in different measurement campaigns. The DTAG-PE05, DTAG-PE06 and DTAG-YSTY cable measurement data is from Deutsche Telekom [4]. The DTAG-PE04 measurement is from the HAInet [9] and the FlexDP research project [5, 10]. The measurement of the HSA-YSTY cable is from a joint work with University of Applied Sciences Augsburg [6, 11]. The British Telecom measurement data for the BT CAD55 cable is published in [12, 13].

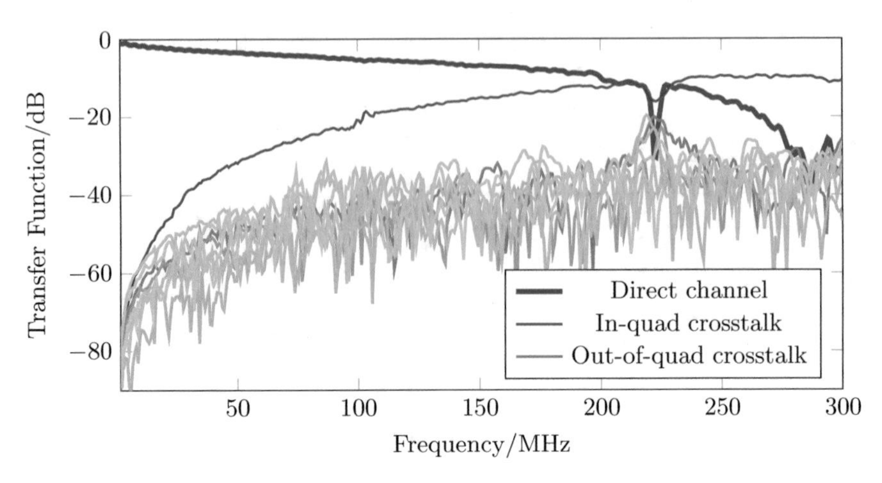

Fig. A.1 Transfer function of direct channel and crosstalk couplings of 30 m DTAG-PE05 line

Figure A.1 shows the direct channel and 9 crosstalk transfer function measurements of one twisted pair from a 10-pair quad-structured DTAG-PE05 cable with 30 m length. The measurement is done up to 300 MHz from Deutsche Telekom [4]. The precise cable name is "I-O2YS(ST)H 10x2x0.5 STVI Bd" as it is described in the TR-285 standard [2]. The cable shows a single dominant crosstalk path which is the crosstalk between two twisted pairs forming a star quad. Otherwise, the crosstalk in this cable is comparably low. This cable is used in Germany as an underground telephone cable, but also for in-building wiring.

Figure A.2 shows a measurement of direct channel and far-end crosstalk for one line of a 10-pair DTAG-PE06 cable binder with 20 m length. The measurement is done up to 300 MHz from Deutsche Telekom [4]. This cable is part of the TR-285 standard as "J-H(ST)H Bd 10x2x0.6". It is a quad-structured cable. But no dominant in-quad crosstalk can be observed. However, the attenuation is higher compared to Fig. A.1 and the out-of-quad crosstalk is higher than at the DTAG-PE05 cable. It is known that the DTAG-PE06 cable is manufactured using alternating twisting while the DTAG-PE05 is manufactured using continuous twisting, which is a possible explanation for the different crosstalk behavior of both cables. This cable is used in Germany as an underground cable and for in-home telephone wiring.

Figure A.3 shows a measurement of a DTAG-YSTY cable with 30 m length with the direct channel and 7 crosstalk paths of the 8-pair cable. The measurement is done up to 300 MHz from Deutsche Telekom [4]. This cable is widely used for in-house wiring. It shows a significantly higher direct channel attenuation than the DTAG-PE05 and DTAG-PE06 cables in Figs. A.1 and A.2, which are the recommended cables from Deutsche Telekom. The main reason for this behavior is that the insulation material of the DTAG-YSTY is polyvinyl chloride while DTAG-PE05 and DTAG-PE06 use polyethylene insulation.

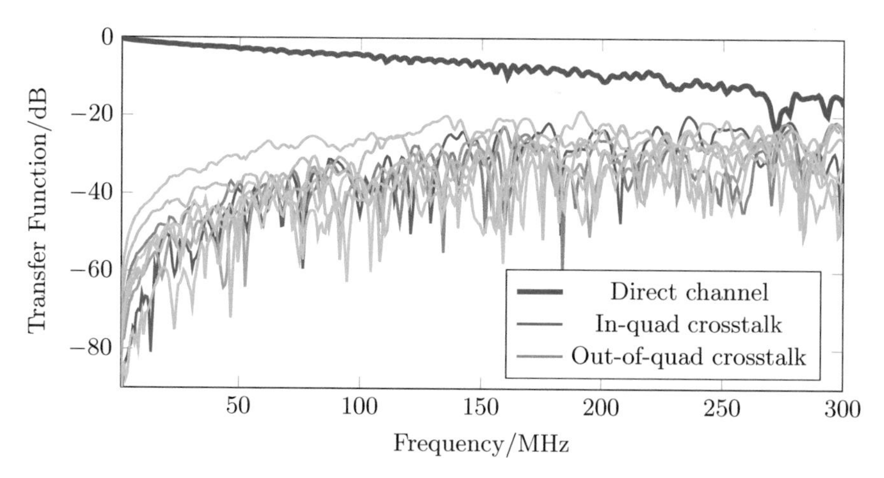

Fig. A.2 Transfer function of direct channel and crosstalk of 20 m DTAG-PE06 line

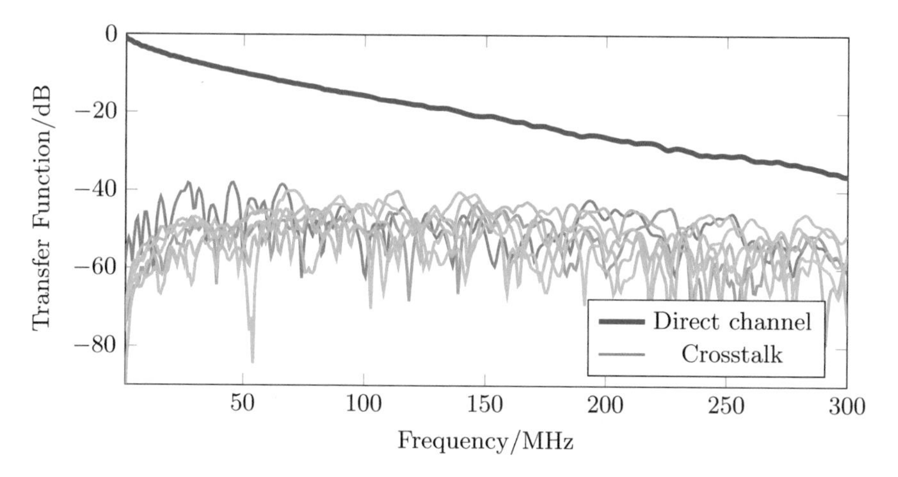

Fig. A.3 Transfer function of direct channel and crosstalk of 30 m DTAG-YSTY line

The DTAG-YSTY cables show a different high frequency behavior, depending on the cable source. Figure A.4 shows measurement data of another cable of the same type from a measurement at Hochschule Augsburg [6]. The measured cable is 20 m long and consists of four twisted pairs. The cable shows a lower attenuation than the measurement from Fig. A.3, but also a very high crosstalk level.

Figure A.5 shows measurement data of a telephone cable used in the British access network [12, 13]. The measured cable is 100 m long and consists of 20 pairs. One direct channel and 9 of the crosstalk paths are shown. The measurement is performed up to 106 MHz.

The cable characteristics are comparable to the DTAG-PE06, but it shows a higher attenuation. This cable is part of [1, 2] with the name CAD55. The cable insulations

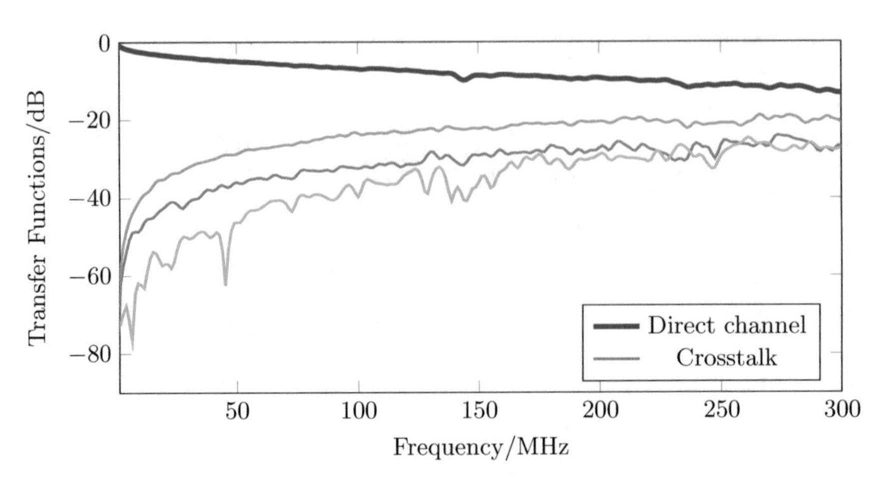

Fig. A.4 Transfer function of direct channel and crosstalk of 20 m HSA-YSTY line

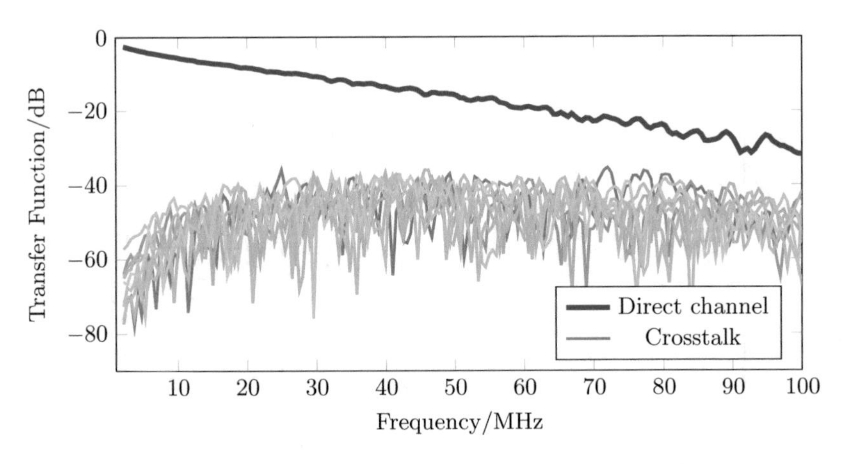

Fig. A.5 Transfer function of direct channel and crosstalk of 100 m BT-CAD55 line

is made of polyethylene, same as for the DTAG-PE05 and the DTAG-PE06. It is a cable with 0.5 mm copper diameter, same as the DTAG-PE05, but showing a higher attenuation than the DTAG-PE05.

A.5 Primary Line Parameter Extraction and Fitting

This section shows the results of the parameter extraction from measurement data on the example of the 8-pair DTAG-YSTY and the 4-pair HSA-YSTY cable. The direct channel primary line parameter measurement is based on an open-short measurement as described in Sect. 2.4.1.

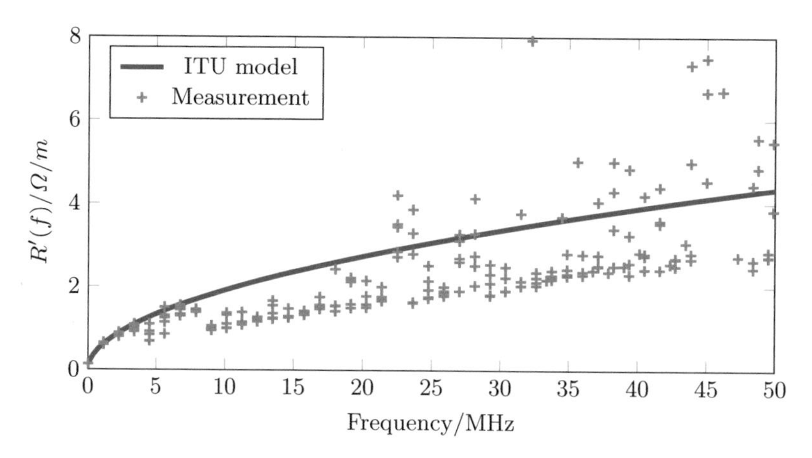

Fig. A.6 Comparison between measured resistance per unit length $R'(f)$ and ITU model

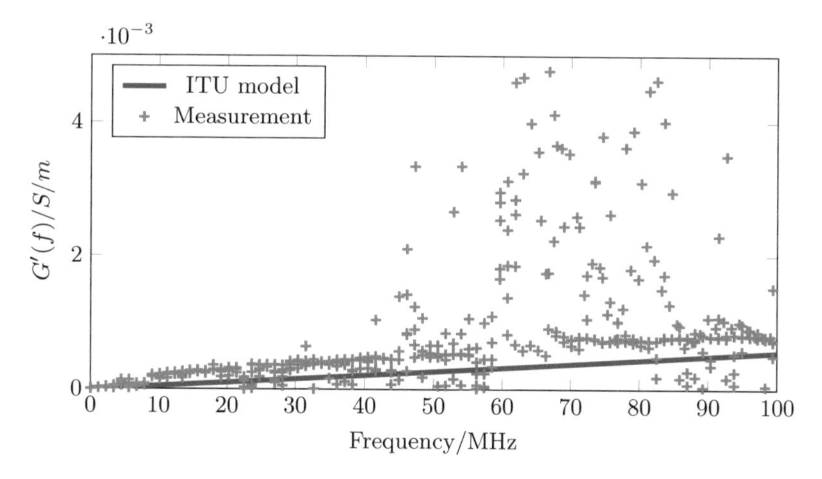

Fig. A.7 Comparison between measured conductance per unit length $G'(f)$ and ITU model

Starting with the primary line parameter extraction, Fig. A.6 compares the result of the parameter fit for the ITU model with measurement data. While the measurement is performed up to 300 MHz, the extraction of resistance per unit length only gives good results at lower frequencies as shown in Fig. A.6. The quality of the parameter fit for the loss terms $R'(f)$ and $G'(f)$ can be verified with the loss term $\mathrm{Re}\{\gamma(f)\}$. A similar behavior is observed for the conductance $G'(f)$ as shown in Fig. A.7.

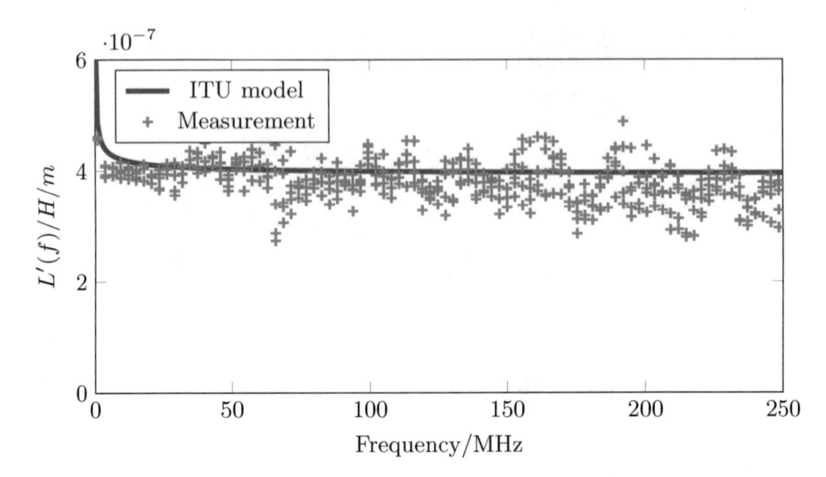

Fig. A.8 Comparison between measured inductance per unit length $L'(f)$ and ITU model

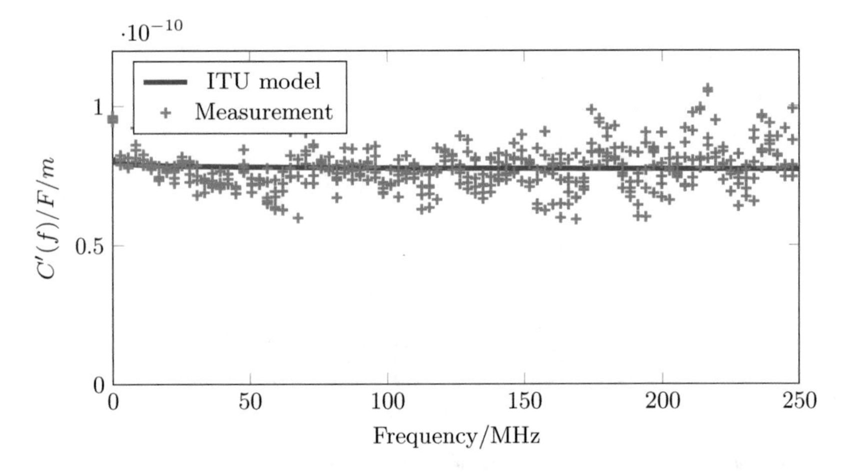

Fig. A.9 Comparison between measured capacitance per unit length $C'(f)$ and ITU model

The parameter extraction for the inductance per unit length $L'(f)$ and the capacitance per unit length $C'(f)$ are shown in Figs. A.8 and A.9, respectively. The primary line parameters, as shown in Figs. A.6, A.7, A.8 and A.9 are the direct result of the ITU model, but they are derived from measurement data with the intermediate step of the secondary line parameters.

The impedance $Z_0(f)$ is shown in Fig. A.10 for the real and Fig. A.11 for the imaginary component. While $\mathrm{Re}\{Z_0(f)\}$ converges from higher values to the line impedance $Z_{0\infty}$, the imaginary component $\mathrm{Im}\{Z_0(f)\}$ converges from negative values to zero. As Fig. A.11 indicates, the extraction of the imaginary component of the line impedance is of low quality and cannot be used for parameter extraction.

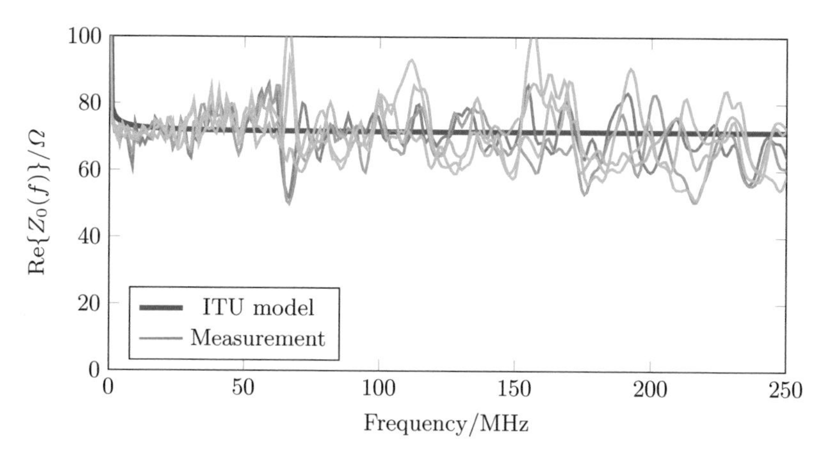

Fig. A.10 Comparison between measured impedance $\mathrm{Re}\{Z_0(f)\}$ and ITU model

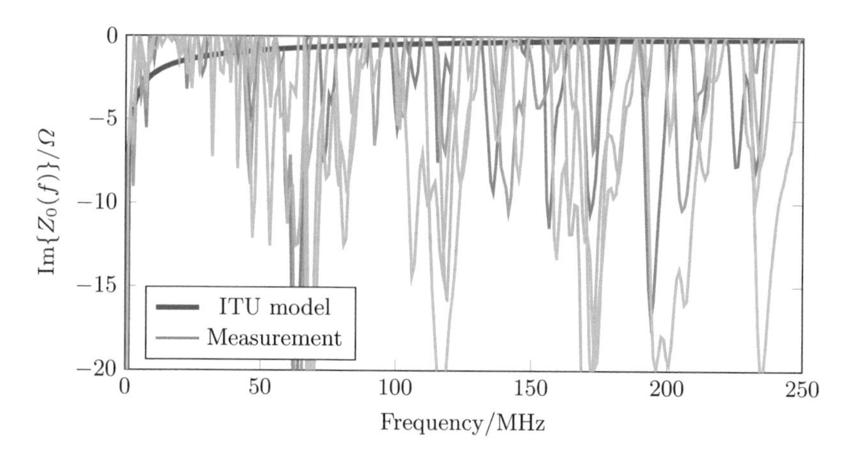

Fig. A.11 Comparison between measured impedance $\mathrm{Im}\{Z_0(f)\}$ and ITU model

The propagation term $\gamma(f)$ is shown in Fig. A.12 for the real component and Fig. A.13 for the imaginary component. A good match between measurement and model is important for the propagation term, because the transfer functions $H(f)$ mainly depends on the propagation term. A comparison between the parameter fit that has been performed for this cable and the measurement indicates a good match between measurement and model.

Finally the line attenuation is compared between the measured and the modeled cable in Fig. A.14. At this point, the direct measurement of the transfer function $H(f)$ is compared with the corresponding model results, which is derived from a model of the primary line parameters after conversion to the secondary line parameters.

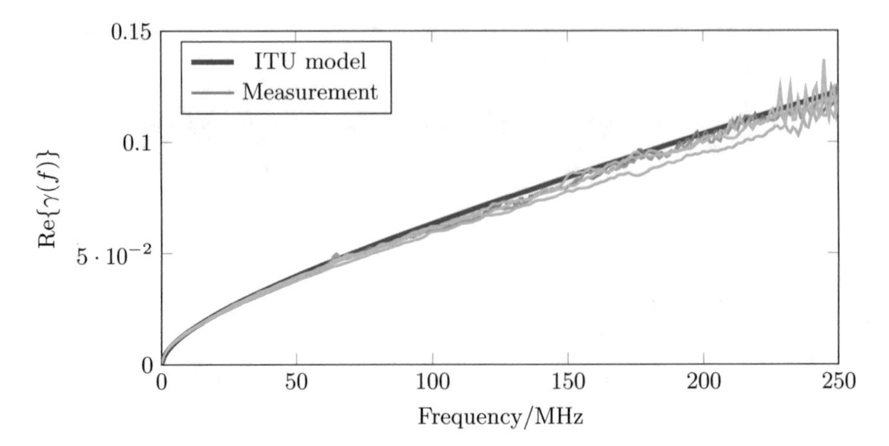

Fig. A.12 Comparison between measured propagation term Re{(f)} and ITU model

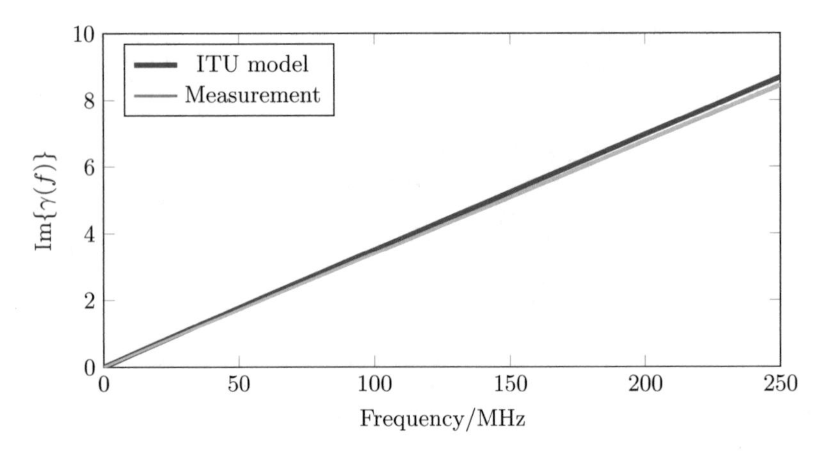

Fig. A.13 Comparison between measured propagation term Im{} and ITU model

The direct channel attenuation shown in Fig. A.14 is besides the crosstalk characteristics the main channel property which determines the achievable data rates on a cable. The comparison shows a good match between measurement and model. However, the direct channel attenuation shows a significant spread between the individual lines of the cable binder. This random component is not covered in the single line models. MIMO modeling approaches are required to describe such behavior.

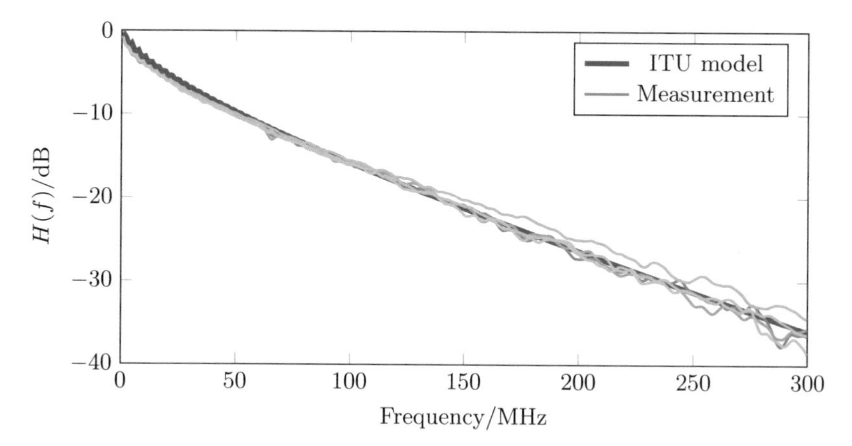

Fig. A.14 Comparison between measured direct channel transfer function and the ITU model

A.6 Example for MIMO Parameter Extraction

The statistics of the $\boldsymbol{X}_{\mathrm{dB}}$-Matrix are one important measure to characterize crosstalk. The DTAG-YSTY cable is used to demonstrate the results. Figure A.15 shows the crosstalk CDF $F(X_{\mathrm{dB}})$ of the measured 8-pair DTAG-YSTY cable for different frequencies in comparison with the crosstalk statistics assumed in the ATIS model [8].

It must be noted that only a small number of measurements is used to perform the presented statistical analysis. For a more detailed analysis, more measurement data is required. Besides the statistics of the crosstalk transfer functions, the inductance matrix is an important figure to compare the channel model with measured cables.

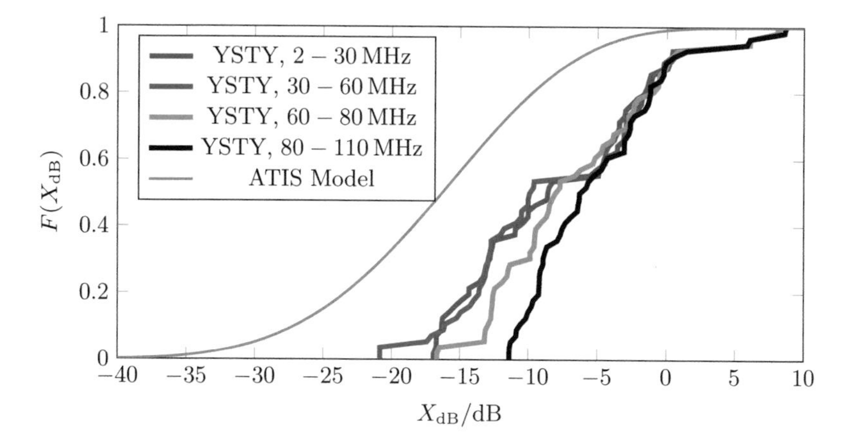

Fig. A.15 Measured crosstalk CDF of the YSTY cable in comparison with the ATIS model [8]

$$\begin{matrix}
0.81 & 0.24 & 0.34 & 0.34 & 0.18 & 0.17 & 0.18 & 0.18 \\
0.24 & 0.79 & 0.34 & 0.34 & 0.18 & 0.19 & 0.18 & 0.18 \\
0.34 & 0.34 & 0.77 & 0.24 & 0.18 & 0.18 & 0.18 & 0.17 \\
0.34 & 0.34 & 0.24 & 0.83 & 0.18 & 0.18 & 0.18 & 0.18 \\
0.18 & 0.17 & 0.18 & 0.18 & 0.79 & 0.23 & 0.33 & 0.34 \\
0.17 & 0.18 & 0.17 & 0.18 & 0.23 & 0.80 & 0.33 & 0.34 \\
0.18 & 0.18 & 0.18 & 0.17 & 0.34 & 0.33 & 0.80 & 0.23 \\
0.18 & 0.18 & 0.17 & 0.18 & 0.34 & 0.33 & 0.23 & 0.79
\end{matrix}$$

Fig. A.16 Measured reference inductance matrix of a 4-quad Y(St)Y-Cable

$$\begin{matrix}
0.81 & 0.24 & 0.33 & 0.33 & 0.15 & 0.18 & 0.16 & 0.18 \\
0.25 & 0.74 & 0.34 & 0.33 & 0.17 & 0.14 & 0.19 & 0.17 \\
0.33 & 0.34 & 0.82 & 0.23 & 0.18 & 0.16 & 0.15 & 0.18 \\
0.34 & 0.34 & 0.24 & 0.73 & 0.15 & 0.18 & 0.18 & 0.17 \\
0.15 & 0.17 & 0.17 & 0.15 & 0.74 & 0.23 & 0.34 & 0.34 \\
0.17 & 0.15 & 0.15 & 0.17 & 0.23 & 0.81 & 0.32 & 0.34 \\
0.15 & 0.18 & 0.17 & 0.17 & 0.32 & 0.34 & 0.74 & 0.23 \\
0.17 & 0.16 & 0.17 & 0.16 & 0.33 & 0.33 & 0.23 & 0.81
\end{matrix}$$

Fig. A.17 Modeled reference inductance matrix of a 4-quad Y(St)Y-Cable

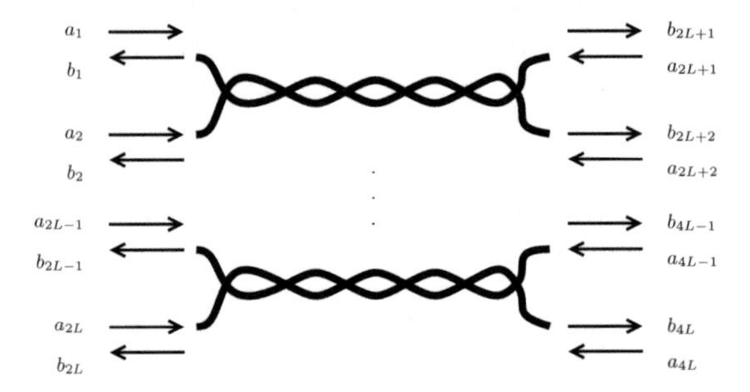

Fig. A.18 Port numbers for multiconductor scattering parameters of multiple twisted pairs

The spatial domain MIMO model uses a single-ended description of the inductance matrix. Therefore, a single-ended measurement of the 4-pair Y(St)Y-Cable is used as reference. It is referred to as HSA-YSTY cable. Figure A.4 shows the corresponding transfer functions.

Figures A.16 and A.17 compare the measured and extracted reference inductance matrices for the HSA-YSTY cable.

A.7 Matrix Definitions for Multiconductor Transmission Lines

The chain matrix description for the differential mode on a single twisted pair is described by

$$\begin{bmatrix} U(0) \\ I(0) \end{bmatrix} = \boldsymbol{A} \begin{bmatrix} U(z) \\ I(z) \end{bmatrix} \tag{A.11}$$

where the chain matrix $\boldsymbol{A}$ is given by

$$\boldsymbol{A} = \begin{bmatrix} \cosh(\gamma z) & Z_0 \sinh(\gamma z) \\ \frac{1}{Z_0} \sinh(\gamma z) & \cosh(\gamma z). \end{bmatrix} \tag{A.12}$$

This refers to the circuit according to Fig. A.19.

Extending the model to multiple twisted pair lines gives the circuit of Fig. A.20. Accordingly, the chain matrix description is extended to be

$$\begin{bmatrix} U_1(0) \\ \vdots \\ U_{2L}(0) \\ I_1(0) \\ \vdots \\ I_{2L}(0) \end{bmatrix} = \boldsymbol{A} \begin{bmatrix} U_1(z) \\ \vdots \\ U_{2L}(z) \\ I_1(z) \\ \vdots \\ I_{2L}(z) \end{bmatrix} \tag{A.13}$$

The voltages can be collected in voltage vectors $\boldsymbol{u}(z)$ and the currents in the vector $\boldsymbol{i}(z)$. The chain matrix is partitioned in four block matrices $\boldsymbol{A}_{11}$ to $\boldsymbol{A}_{22}$ to give the chain matrix description

$$\begin{bmatrix} \boldsymbol{u}(0) \\ \boldsymbol{i}(0) \end{bmatrix} = \begin{bmatrix} \boldsymbol{A}_{11} & \boldsymbol{A}_{12} \\ \boldsymbol{A}_{21} & \boldsymbol{A}_{22} \end{bmatrix} \cdot \begin{bmatrix} \boldsymbol{u}(z) \\ \boldsymbol{i}(z) \end{bmatrix} \tag{A.14}$$

This description allows to describe cascades of multi-port circuits such as multiple segments of a cable bundle.

While the description in voltages and currents is convenient for the analysis of the equivalent circuit of the transmission line, a description of transmitted and reflected

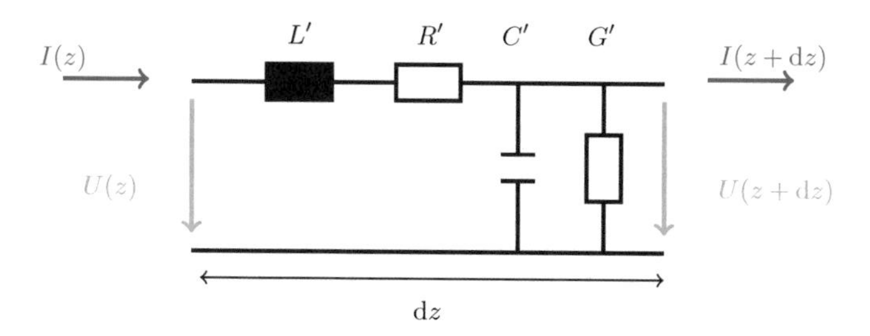

Fig. A.19 Model of a differential line element dz

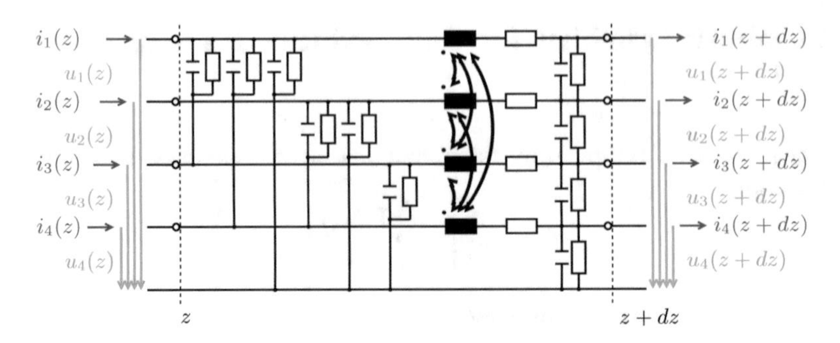

Fig. A.20 Model of a binder element dz for two twisted pairs

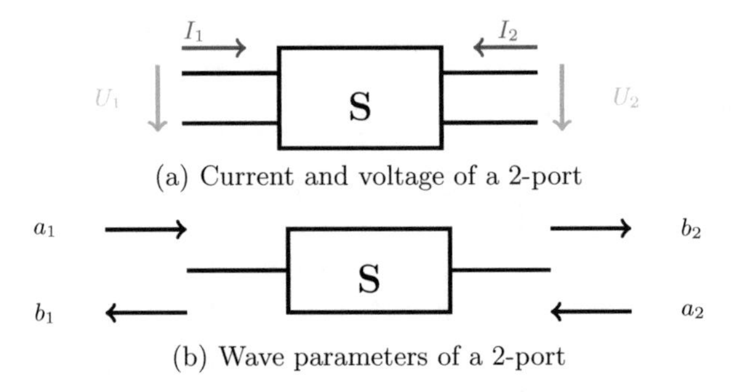

Fig. A.21 2-port chain matrix and 2-port scattering matrix

waves is more suitable for signal transmission models. The two-port network of Fig. A.19 is equivalent to a two-port with the incident waves a_1 and a_2 and the reflected waves b_1 and b_2. They are derived from voltage and current according to

$$
\begin{aligned}
a_1 &= \frac{U_1}{\sqrt{Z_0}} + I_1 \cdot \sqrt{Z_0} \\
a_2 &= \frac{U_2}{\sqrt{Z_0}} + I_2 \cdot \sqrt{Z_0} \\
b_1 &= \frac{U_1}{\sqrt{Z_0}} - I_1 \cdot \sqrt{Z_0} \\
b_2 &= \frac{U_2}{\sqrt{Z_0}} - I_2 \cdot \sqrt{Z_0}
\end{aligned}
\tag{A.15}
$$

with a reference impedance Z_0. The 2-port is described by the scattering matrix $\boldsymbol{S}$ as shown in Fig. A.21. Voltages and currents of Fig. A.21a are translated to waves in Fig. A.21b.

Again, the description is extended to multiple ports according to Fig. A.22.

The dependency between incident waves and reflected waves of the two-port is given by

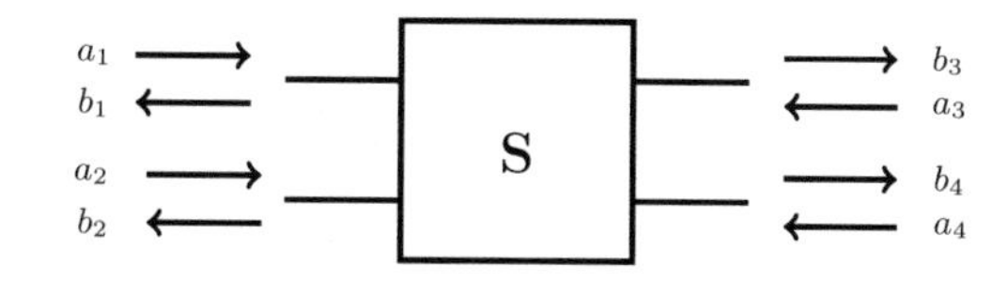

Fig. A.22 4-port scattering matrix

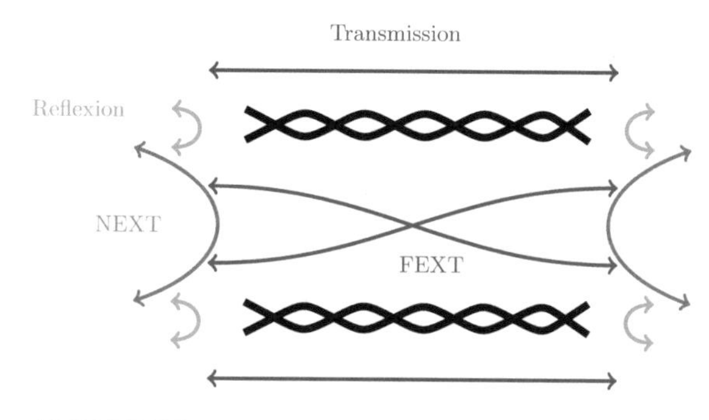

Fig. A.23 Transmission, reflection, FEXT and NEXT of two lines

Table A.7 Physical effects and the corresponding scattering parameters of two twisted-pair lines

Effect	Scattering parameter
Transmission	S_{13}, S_{24}, S_{31} und S_{42}
Reflection	S_{11}, S_{22}, S_{33} und S_{44}
NEXT	S_{12}, S_{34}, S_{21} und S_{43}
FEXT	S_{14}, S_{41}, S_{32} und S_{23}

$$\begin{bmatrix} b_1 \\ b_2 \end{bmatrix} = \begin{bmatrix} S_{11} & S_{12} \\ S_{21} & S_{22} \end{bmatrix} \cdot \begin{bmatrix} a_1 \\ a_2 \end{bmatrix}. \tag{A.16}$$

A multi-port according to Fig. A.22 is partitioned according to

$$\begin{bmatrix} b_1 \\ \vdots \\ b_L \\ b_{L+1} \\ \vdots \\ b_{2L} \end{bmatrix} = \begin{bmatrix} \boldsymbol{S}_{11} & \boldsymbol{S}_{12} \\ \boldsymbol{S}_{21} & \boldsymbol{S}_{22} \end{bmatrix} \cdot \begin{bmatrix} a_1 \\ \vdots \\ a_L \\ a_{L+1} \\ \vdots \\ a_{2L} \end{bmatrix} \tag{A.17}$$

The scattering parameters of the 4-port directly describe the transfer functions of transmission, reflection, far-end and near-end crosstalk as shown in Table A.7 and Fig. A.23.

A.8 Conversion Between Scattering and Chain Matrices

The MIMO cable model is formulated in the domain of chain matrices $\boldsymbol{A}(f)$, while for signal transmission, the scattering matrix $\boldsymbol{S}(f)$ is of interest, as it describes

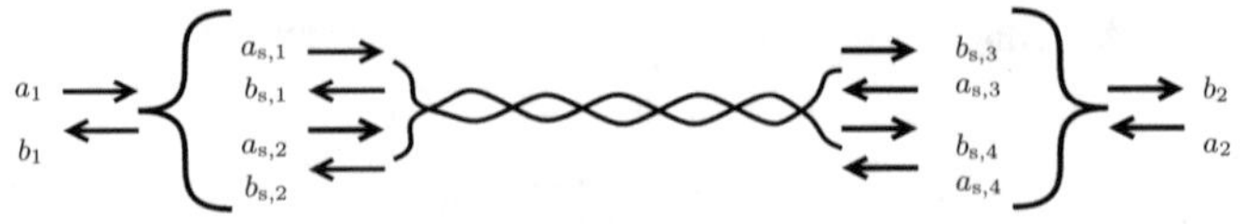

Fig. A.24 Single-ended modes and differential modes of a twisted pair

the direct channel and crosstalk transfer functions. The conversion rules, which are summarized here, can be found in [14, 15].

The conversion of a chain matrix into a scattering matrix is given by

$$\begin{aligned} \boldsymbol{S}_{11} &= (\boldsymbol{A}_{11} + \boldsymbol{A}_{12} - \boldsymbol{A}_{21} - \boldsymbol{A}_{22}) \cdot (\boldsymbol{A}_{11} + \boldsymbol{A}_{12} + \boldsymbol{A}_{21} + \boldsymbol{A}_{22})^{-1} \\ \boldsymbol{S}_{12} &= (\boldsymbol{A}_{11} - \boldsymbol{A}_{21}) - (\boldsymbol{A}_{11} + \boldsymbol{A}_{12} - \boldsymbol{A}_{21} - \boldsymbol{A}_{22}) \\ &\quad \cdot (\boldsymbol{A}_{11} + \boldsymbol{A}_{12} + \boldsymbol{A}_{21} + \boldsymbol{A}_{22})^{-1} \cdot (\boldsymbol{A}_{11} + \boldsymbol{A}_{21}) \\ \boldsymbol{S}_{21} &= 2 \cdot (\boldsymbol{A}_{11} + \boldsymbol{A}_{12} + \boldsymbol{A}_{21} + \boldsymbol{A}_{22})^{-1} \\ \boldsymbol{S}_{22} &= 2 \cdot (\boldsymbol{A}_{11} + \boldsymbol{A}_{12} + \boldsymbol{A}_{21} + \boldsymbol{A}_{22})^{-1} \cdot (-\boldsymbol{A}_{11} - \boldsymbol{A}_{21}). \end{aligned} \tag{A.18}$$

The conversion from a scattering matrix into a chain matrix is given by

$$\begin{aligned} \boldsymbol{A}_{11} &= \frac{1}{2} \left((\boldsymbol{I} + \boldsymbol{S}_{11}) \, \boldsymbol{S}_{21}^{-1} \, (\boldsymbol{I} - \boldsymbol{S}_{22}) + \boldsymbol{S}_{12} \right) \\ \boldsymbol{A}_{12} &= \frac{Z_0}{2} \left((\boldsymbol{I} + \boldsymbol{S}_{11}) \, \boldsymbol{S}_{21}^{-1} \, (\boldsymbol{I} + \boldsymbol{S}_{22}) - \boldsymbol{S}_{12} \right) \\ \boldsymbol{A}_{21} &= \frac{Z_0}{2} \left((\boldsymbol{I} - \boldsymbol{S}_{11}) \, \boldsymbol{S}_{21}^{-1} \, (\boldsymbol{I} - \boldsymbol{S}_{22}) - \boldsymbol{S}_{12} \right) \\ \boldsymbol{A}_{22} &= \frac{1}{2} \left((\boldsymbol{I} - \boldsymbol{S}_{11}) \, \boldsymbol{S}_{21}^{-1} \, (\boldsymbol{I} - \boldsymbol{S}_{22}) + \boldsymbol{S}_{12} \right). \end{aligned} \tag{A.19}$$

A.9 Single-Ended and Differential Mode

The MIMO cable binder model in Sect. 2.2.1 describes the single-ended modes of a twisted pair cable. For data transmission, the differential modes are used.

With reference to twisted pair cables, as shown in Fig. A.24, the single-ended mode describes the waves on individual wires $a_{s,1}$ to $a_{s,4L}$ and $b_{s,1}$ to $b_{s,4L}$ of the twisted pair with respect to a common ground potential. The differential mode describes the differential waves a_1 to a_{2L} and b_1 to b_{2L} between the wires of the twisted pair and the common mode is the common wave of two wires with respect to the ground potential.

The differential mode scattering matrix $\boldsymbol{S}_{\mathrm{dd}}$, which contains the transfer functions and crosstalk according to Table A.7 is derived from the single-ended scattering matrix by the following conversion

$$S_{\text{mm}} = \begin{bmatrix} S_{\text{dd}} & S_{\text{dc}} \\ S_{\text{cd}} & S_{\text{cc}} \end{bmatrix} = \frac{1}{2} M S M_{\text{inv}} \tag{A.20}$$

where the matrix M is given by

$$M = \begin{bmatrix} 1 & -1 & 0 & 0 & \dots & & 0 \\ 0 & 0 & 1 & -1 & \dots & & 0 \\ \vdots & & \ddots & & \ddots & & \vdots \\ 0 & & & \dots & & 1 & -1 \\ 1 & 1 & 0 & 0 & \dots & & 0 \\ 0 & 0 & 1 & 1 & \dots & & 0 \\ \vdots & & \ddots & & \ddots & & \vdots \\ 0 & & & \dots & & 1 & 1 \end{bmatrix} \tag{A.21}$$

and the matrix M_{inv} is given by

$$M_{\text{inv}} = \begin{bmatrix} 1 & 0 & \dots & 0 & 1 & 0 & \dots & 0 \\ -1 & 0 & & 0 & 1 & 0 & & 0 \\ 0 & 1 & & & 0 & 1 & & \\ 0 & -1 & \ddots & 0 & 0 & 1 & \ddots & 0 \\ \vdots & & \ddots & 0 & \vdots & & \ddots & 0 \\ 0 & & & 1 & 0 & & & 1 \\ 0 & & \dots & -1 & 0 & & \dots & 1 \end{bmatrix}. \tag{A.22}$$

S_{mm} is the mixed-mode matrix. It contains besides the differential S parameters S_{dd} which are used for data transmission also the common mode matrix S_{cc} and the conversion mode matrices S_{dc} and S_{cd}.

References

1. ITU-T Rec. G.9701: Fast access to subscriber terminals - physical layer specification. ITU Recommendation (2015)
2. Broadband Forum TR-285: Cable models for physical layer testing of G.fast access network. Technical report (2015)
3. Humphrey, L., Morsman, T.: G.fast: release of BT cable measurements for use in simulations (2013). ITU-T Contribution SG15/Q4a 2013-1-Q4-066
4. Muggenthaler, P.: Ermittlung der Leitungsbeläge und ELFEXT(F) verschiedener Installationskabel in der Kabelversuchsanlage der Deutschen Telekom als Grundlage für das G.fast Kanalmodell (2012). Internal report
5. Maierbacher (Fraunhofer ESK), G., Strobel (Lantiq Deutschland GmbH), R., Liss (InnoRoute GmbH), C.: Entwicklung eines Generators für Mustermodelle verkoppelter Mehrleitersysteme fr die FTTdp-Übertragung. Technical report, Bayerische Forschungsstiftung (2014)
6. Blenk, T.: Entwicklung eines Generators für Mustermodelle verkoppelter Mehrleitersysteme für die FTTdp-Übertragung. Master's thesis, Hochschule Augsburg, University of Applied Sciences (2014)

7. ETSI TS 101 270-1: Transmission and multiplexing (TM); Access transmission systems on metallic access cables; Very high speed digital subscriber Line (VDSL); Part 1: functional requirements (2003)
8. Maes, J., Guenach, M., Peeters, M.: Statistical channel model for gain quantification of DSL crosstalk mitigation techniques. In: IEEE International Conference on Communications (ICC)
9. Eder (Lantiq Deutschland GMbH), A., Helfer (CAD Service), G., Leibiger (Fraunhofer ESK), M.: Hochbitratige Access- und Inhausnetze (HAInet) Schlussbericht. Technical report, Forderprogramm IuK Bayern (2014)
10. Maierbacher, G., Strobel, R., Liss, C.: Abschlussbericht FlexDP (AZ-1097-13). Technical report, Bayerische Forschungsstiftung (2016)
11. Blenk, T.: Analyse existierender Leitungsmodelle zur Charakterisierung von DSL-Leitungen. Bachelor's thesis, Hochschule Augsburg, University of Applied Sciences (2014)
12. G.fast: Release of BT cable (20 pair) measurements for use in simulations (2015). ITU-T Contribution SG15/Q4a 2015-02-Q4-053
13. Horsley, I., Singleton, H.: Release of BT cable (20 pair) NEXT measurements for use in simulations (2015). ITU-T Contribution SG15/Q4 COM15 - C1499 - E
14. Granberg, T.: Handbook of Digital Techniques for High-Speed Design. Prentice Hall PTR, Upper Saddle River (2004)
15. Fan, W., Lu, A., Wai, L., Lok, B.: Mixed-mode S-parameter characterization of differential structures. In: Conference on Electronics Packaging Technology (EPTC), pp. 533–537. IEEE (2003)

Appendix B

B.1 Model for Discrete Multi-tone Modulation

This section presents a matrix-based model of discrete multi-tone modulation (DMT), which describes the frequency domain channel as well as intercarrier interference. The model describes the following steps:

1. inverse discrete Fourier transform (IDFT) at the transmitter,
2. generation of cyclic prefix and cyclic suffix,
3. convolution with the channel impulse response,
4. removal of the cyclic extension at the receiver and
5. discrete Fourier transform (DFT).

While OFDM systems translate K complex frequency domain samples into K complex time domain samples, DMT modulation creates a baseband signal with $2K$ real-valued samples in time domain. A standard complex model is not sufficient to describe DMT modulation. Therefore, real and complex matrix components are modeled independently according to [1]

$$\boldsymbol{y} = \boldsymbol{A}\boldsymbol{x} \rightarrow \begin{bmatrix} \mathrm{Re}\{\boldsymbol{y}\} \\ \mathrm{Im}\{\boldsymbol{y}\} \end{bmatrix} = \begin{bmatrix} \mathrm{Re}\{\boldsymbol{A}\} & \mathrm{Im}\{\boldsymbol{A}\} \\ -\mathrm{Im}\{\boldsymbol{A}\} & \mathrm{Re}\{\boldsymbol{A}\} \end{bmatrix} \begin{bmatrix} \mathrm{Re}\{\boldsymbol{x}\} \\ \mathrm{Im}\{\boldsymbol{x}\} \end{bmatrix}. \tag{B.1}$$

The coefficients of inverse discrete Fourier transform matrix $\boldsymbol{F}^{-1} \in \mathbb{C}^{2K \times 2K}$ are given by

$$[\boldsymbol{F}^{-1}]_{ik} = \mathrm{e}^{2\pi \mathrm{j} \cdot \frac{(i-1)(k-1)}{2K}} \tag{B.2}$$

for K carriers. To have real-valued time domain samples and to match the framework of Eq. (B.1), the input samples are multiplied with the matrix $\boldsymbol{C}_{\mathrm{ri}} \in \mathbb{R}^{2K \times 2K}$, which is given by

R. Strobel, *Channel Modeling and Physical Layer Optimization in Copper Line Networks*, Signals and Communication Technology,
https://doi.org/10.1007/978-3-319-91560-9

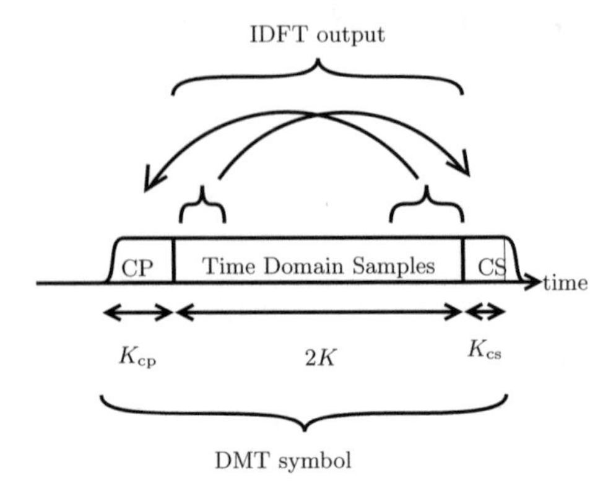

Fig. B.1 Construction of time domain samples of a DMT symbol with cyclic prefix and cyclic suffix

$$\boldsymbol{C}_{\text{ri}} = \begin{bmatrix} & \boldsymbol{0}_{1\times 2K} & & \\ \boldsymbol{0}_{K-1\times 1} & \boldsymbol{I}_{K-1} & \boldsymbol{0}_{K-1\times 1} & j\boldsymbol{I}_{K-1} \\ & \boldsymbol{0}_{1\times 2K} & & \\ \boldsymbol{0}_{K-1\times 1} & \bar{\boldsymbol{I}}_{K-1} & \boldsymbol{0}_{K-1\times 1} & -j\bar{\boldsymbol{I}}_{K-1} \end{bmatrix} \tag{B.3}$$

with $\bar{\boldsymbol{I}}$ to be the anti-identity matrix. The product $\boldsymbol{F}^{-1}\boldsymbol{C}_{\text{ri}}$ performs the real-valued IDFT and forces the carrier $k = 1$ to zero.

The second step after the IDFT is to add the cyclic prefix of K_{cp} samples and the cyclic suffix of K_{cs} samples as shown in Fig. B.1. This is performed by multiplying the matrix $\boldsymbol{C}_{\text{a}} \in \mathbb{R}^{K_{\text{cp}}+2K+K_{\text{cs}}\times 2K}$ which is given by

$$\boldsymbol{C}_{a} = \begin{bmatrix} \boldsymbol{0}_{K_{\text{cp}}\times 2K-K_{\text{cp}}} & \boldsymbol{I}_{K_{\text{cp}}} \\ \boldsymbol{I}_{2K} & \\ \boldsymbol{I}_{K_{\text{cs}}} & \boldsymbol{0}_{K_{\text{cs}}\times 2K-K_{\text{cs}}} \end{bmatrix}. \tag{B.4}$$

The convolution with the channel impulse response $\boldsymbol{h}_{\text{td}} = [h_1, \ldots, h_{K_{\text{ch}}}]^{\text{T}}$ is described with a convolution matrix $\boldsymbol{H}_{\text{td}}$. The full convolution matrix $\boldsymbol{H}_{\text{td,ext}} \in \mathbb{R}^{2K+K_{\text{cp}}+K_{\text{cs}}\times 2K+K_{\text{cp}}+K_{\text{cs}}+K_{\text{ch}}-1}$ is given by

$$\boldsymbol{H}_{\text{td,ext}} = \begin{bmatrix} h_1 & h_2 & \ldots & h_{K_{\text{ch}}} & 0 & 0 & \ldots & 0 \\ 0 & h_1 & \ddots & h_{K_{\text{ch}}-1} & h_{K_{\text{ch}}} & 0 & \ldots & 0 \\ & & & \ddots & \ddots & & & \\ 0 & 0 & \ldots & & & \ldots & h_{K_{\text{ch}}-1} & h_{K_{\text{ch}}} \end{bmatrix}. \tag{B.5}$$

The matrix $\boldsymbol{H}_{\text{td,ext}}$ has more columns than the number of rows of $\boldsymbol{C}_{\text{a}}$. Only the columns $K_{\text{offset}} + 1$ to $K_{\text{offset}} + 2K + K_{\text{cp}} + K_{\text{cs}}$ are taken for the matrix $\boldsymbol{H}_{\text{td}}$. The offset K_{offset} determines the adjustment between transmit and receive timing. For the typical shape of the impulse response of a twisted pair line, as presented in Fig. B.2, it is selected to match the line delay and the length of the cyclic prefix.

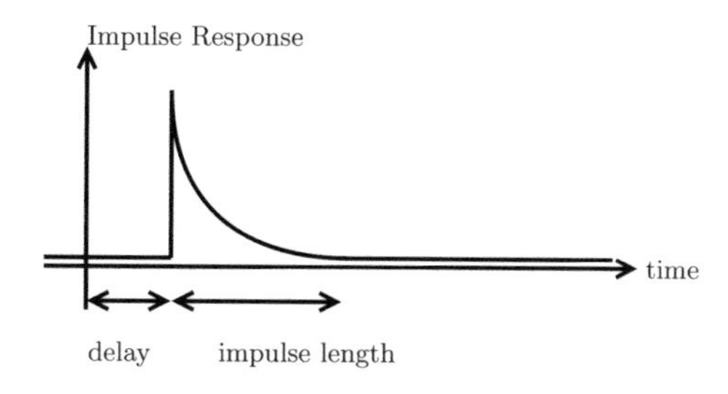

Fig. B.2 Impulse response of twisted pair line

At the receiver side, the cyclic extension is removed, which is described by the matrix $\boldsymbol{C}_{\mathrm{r}} \in \mathbb{R}^{2K \times K_{\mathrm{cp}}+2K+K_{\mathrm{cs}}}$ according to

$$\boldsymbol{C}_r = \left[\boldsymbol{0}_{2K \times K_{\mathrm{cp}}} \ \boldsymbol{I}_{2K} \ \boldsymbol{0}_{2K \times K_{\mathrm{cs}}} \right]. \tag{B.6}$$

DFT is performed over the selected $2K$ samples with the DFT matrix $\boldsymbol{F} \in \mathbb{R}^{2K \times 2K}$. According to the framework of Eq. (B.1), it is given by

$$f_{ik} = \begin{cases} \cos \frac{-2\pi(i-1)(k-1)}{2K} & \text{for } i \leq K \\ \sin \frac{-2\pi(i-1-K)(k-1)}{2K} & \text{for } i > K. \end{cases} \tag{B.7}$$

The matrix model of DMT modulation is given by

$$\boldsymbol{H}_{\mathrm{fd}} = \boldsymbol{F} \cdot \boldsymbol{C}_{\mathrm{r}} \boldsymbol{H}_{\mathrm{td}} \boldsymbol{C}_{\mathrm{a}} \boldsymbol{F}^{-1} \boldsymbol{C}_{\mathrm{ri}}. \tag{B.8}$$

With sufficient length of the cyclic extension, $K_{\mathrm{cp}} + K_{\mathrm{cs}} \geq K_{\mathrm{ch}} - 1$, there is no intercarrier interference and $\boldsymbol{H}_{\mathrm{fd}}$ consists of diagonal matrices. In this case, the channel on carrier k is represented by a single coefficient, which is

$$H^{(k)} = [\boldsymbol{H}_{\mathrm{fd}}]_{k,k} + j[\boldsymbol{H}_{\mathrm{fd}}]_{k,k+K}. \tag{B.9}$$

The intercarrier interference which is created when $K_{\mathrm{cp}} + K_{\mathrm{cs}} < K_{\mathrm{ch}} - 1$ holds, is not circularly symmetric.

B.2 SNR-Dependent BER for Uncoded QAM Modulation

Channel capacity is given by $C = \log_2(1 + SNR)$, while for the achievable data rate, the approximation $b = \log_2\left(1 + \frac{SNR}{\Gamma}\right)$ with the SNR gap Γ is used [2]. The SNR gap for the coding and modulation scheme used in DSL is often approximated with Γ= 9.8 dB [2], but the actual gap to capacity depends on the constellation size and the modulation settings.

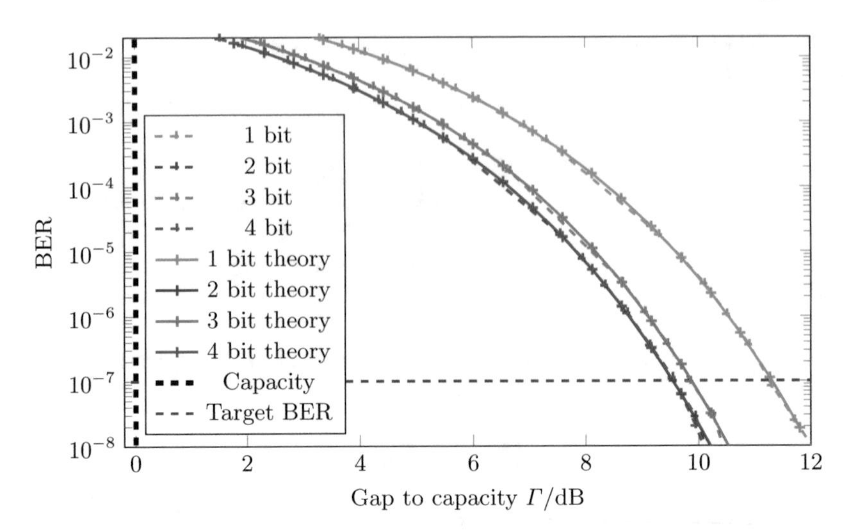

Fig. B.3 Comparison between theoretical uncoded BER and simulated BER for linear receivers up to 4 bits

Figures B.3 and B.4 show the comparison between theoretical values for the bit error rate, derived from a lower bound on the uncoded QAM bit error rate with the simulated bit error rate, using a hardware model of the G.fast receiver. The SNR gap Γ to capacity is shown for different constellations. The results show that the simulated bit error rate is very close to the lower bound, while the approximation with $\Gamma = 9.8$ dB for all constellations does not always give a good fit.

The bit error rate lower bound is derived as follows. Assuming a Gaussian channel, where the desired signal u consists of the real part $u_{\text{re}} = \text{Re}\{\text{u}\}$ and the imaginary part $u_{\text{im}} = \text{Im}\{\text{u}\}$, the receive signal with additive white Gaussian noise is $\hat{u}_{\text{re/im}} \sim \mathcal{N}\left(u_{\text{re/im}}, \frac{|g|^2\sigma^2}{2}\right)$.

The probability of a decision error towards the neighboring constellation point with a distance d according to Fig. B.5 is obtained by integration over the area of the receive signal probability density function $f_{\hat{u}}(\hat{u}) = f_{\hat{u}_{\text{re}}}(\hat{u}_{\text{re}}) f_{\hat{u}_{\text{im}}}(\hat{u}_{\text{im}})$ causing the corresponding decision error.

For the one bit constellation, the probability of a decision error is given by

$$p_{\text{flip},1} = \int\limits_{\hat{u}_{\text{re}}=-\infty}^{u_{\text{re}}-d/2} f_{\hat{u}_{\text{re}}}(\hat{u}_{\text{re}}) \mathrm{d}\hat{u}_{\text{re}} = F_{\hat{u}_{\text{re}}}(u_{\text{re}} - d/2), \tag{B.10}$$

where $F_{\hat{u}_{\text{re}}}(\hat{u}_{\text{re}})$ is the cumulative density function of the receive signal real or imaginary part, assuming that $f_{\hat{u}_{\text{re}}}(\hat{u}_{\text{re}}) = f_{\hat{u}_{\text{im}}}(\hat{u}_{\text{im}})$ holds for the noisy receive signal.

For constellations with two or more bits, a symbol error may cause multiple bit flips. The 8 directly neighboring constellation points are considered to derive the bit

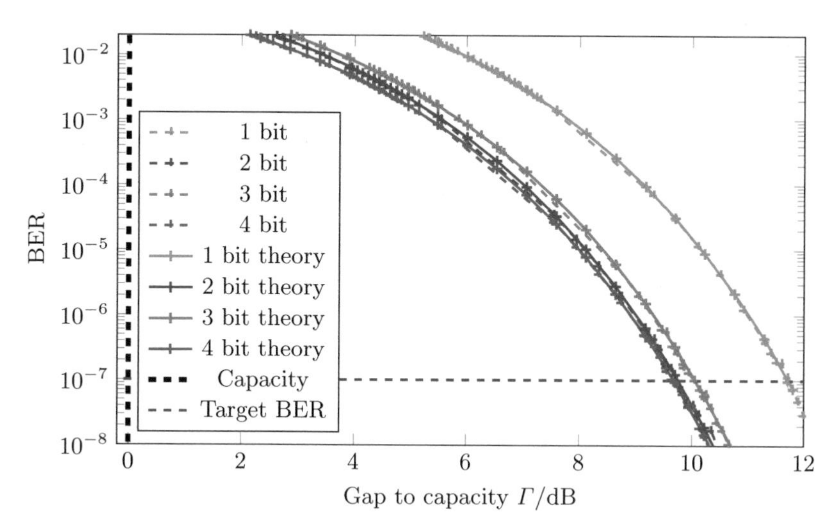

Fig. B.4 Comparison between theoretical uncoded BER and simulated BER for modulo receivers up to 4 bits

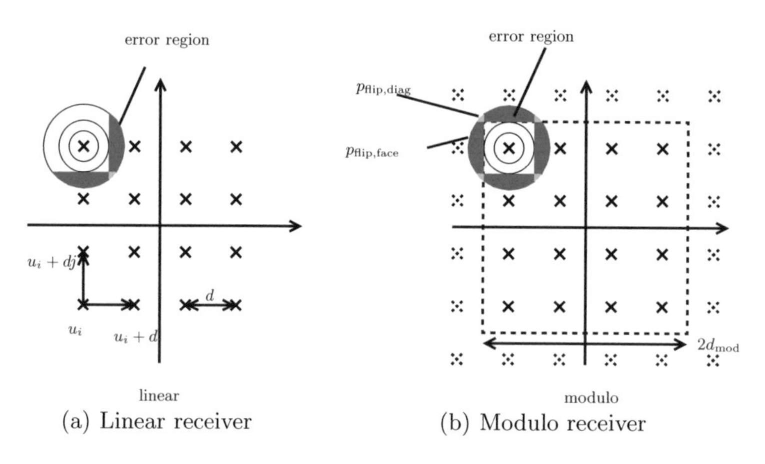

Fig. B.5 Comparison of error regions

error probability lower bound. As indicated in Fig. B.5b, there are two different error probabilities. $p_{\text{flip,face}}$ for the four closest points

$$p_{\text{flip,face}} = \iint\limits_{\substack{\hat{u}_{\text{re}}=u_{\text{re}}-3d/2 \\ \hat{u}_{\text{im}}=u_{\text{im}}-d/2}}^{\substack{u_{\text{re}}-d/2 \\ u_{\text{im}}+d/2}} f_{\hat{u}}(\hat{u}) \mathrm{d}\hat{u}_{\text{re}} \mathrm{d}\hat{u}_{\text{im}} \approx p_{\text{flip},1} - 2p_{\text{flip},1}^2 \tag{B.11}$$

Table B.1 Average number of erroneous bits per symbol error

$\hat{b}$	$n_{\text{face,lin}}$	$n_{\text{diag,lin}}$	$n_{\text{face,mod}}$	$n_{\text{diag,mod}}$
1	0.2500	0.5000	1.0000	0
2	0.5000	1.0000	1.0000	2.0000
3	0.8125	1.6250	1.5000	2.5000
4	1.0000	2.0000	1.5000	3.0000
5	1.5312	2.7656	1.6562	3.0469
6	1.3750	2.7500	1.7500	3.5000
7	1.6641	3.2539	1.7891	3.5039
8	1.6250	3.2500	1.8750	3.7500
9	1.7793	3.5381	1.8730	3.7256
10	1.7812	3.5625	1.9375	3.8750
11	1.8638	3.7219	1.9263	3.8469
12	1.8750	3.7500	1.9688	3.9375

Table B.2 SNR versus bit allocation table for uncoded QAM modulation, linear receiver

Bits	1	2	3	4	5	6	7
SNR	11.35	14.33	18.31	21.30	24.33	27.58	30.48
Bits	8	9	10	11	12	13	14
SNR	33.59	36.37	39.60	42.39	45.53	48.43	51.53

and $p_{\text{flip,diag}}$ (see Fig. B.5b) for the diagonally neighboring constellation points according to[1]

$$p_{\text{flip,diag}} = \int\limits_{\substack{\hat{u}_{\text{re}}=u_{\text{re}}-3d/2 \\ \hat{u}_{\text{im}}=u_{\text{im}}-3d/2}}^{\substack{u_{\text{re}}-d/2 \\ u_{\text{im}}-d/2}} \int f_{\hat{u}}(\hat{u}) \mathrm{d}\hat{u}_{\text{re}} \mathrm{d}\hat{u}_{\text{im}} \approx p_{\text{flip,1}}^2. \tag{B.12}$$

Comparing Fig. B.5a with b, the number of possible bit errors is different between the linear and modulo receiver because for the modulo receiver, constellation points may repeat outside of the square modulo region. Counting the average number of erroneous bits for a symbol error for the two different error probabilities $p_{\text{flip,face}}$ and $p_{\text{flip,diag}}$ gives the corresponding number of erroneous bits n_{face} and n_{diag} according to Table B.1.

From Eqs. (B.11), (B.12) and Table B.1, the uncoded bit error rate $p_{\text{error},\hat{b}}$ is

[1]The exact solutions for the integrals of Eqs. (B.11) and (B.12) would be $p_{\text{flip,face}} = p_{\text{flip,1}} - 2p_{\text{flip,1}}^2 - (p_{\text{flip,2}} - 2p_{\text{flip,2}}p_{\text{flip,1}})$ and $p_{\text{flip,diag}} = p_{\text{flip,1}}^2 - (p_{\text{flip,2}}^2 - 2p_{\text{flip,2}}p_{\text{flip,1}})$ with $p_{\text{flip,2}} = F_{\hat{u}_{\text{re}}}(u_{\text{re}} - 3d/2)$, but for the SNR region of interest, the difference between and the precise lower bounds can be ignored.

Table B.3 SNR versus bit allocation table for uncoded QAM modulation, modulo receiver

Bits	1	2	3	4	5	6	7
SNR	11.70	14.50	18.50	21.42	24.42	27.55	30.45
Bits	8	9	10	11	12	13	14
SNR	33.60	36.40	39.60	42.39	45.55	48.33	51.59

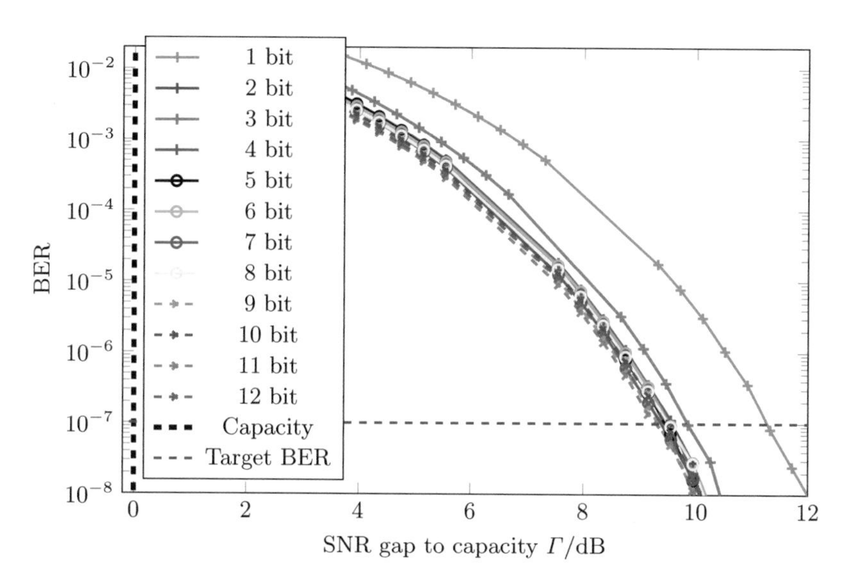

Fig. B.6 BER versus SNR curves for linear receivers, uncoded QAM modulation, based on a hardware model of the G.fast receiver

$$p_{\text{error},\hat{b}} = \frac{4n_{\text{face},\hat{b}}p_{\text{flip,face}} + 4n_{\text{diag},\hat{b}}p_{\text{flip,diag}}}{\hat{b}}. \tag{B.13}$$

Conversely, the required SNR $SNR_{\text{req},\hat{b}}$ can be defined as

$$SNR_{\text{req},\hat{b}} = \min SNR \text{ s.t., } p_{\text{error},\hat{b}}(SNR) \leq p_{\text{error,target}} \tag{B.14}$$

The corresponding SNR values to achieve the target bit error rate for each constellation are summarized in Tables B.2 and B.3.

Figures B.6 and B.7 show the simulated bit error rate versus the SNR gap to capacity for G.fast constellations from 1 to 12 bit for linear and modulo receivers.

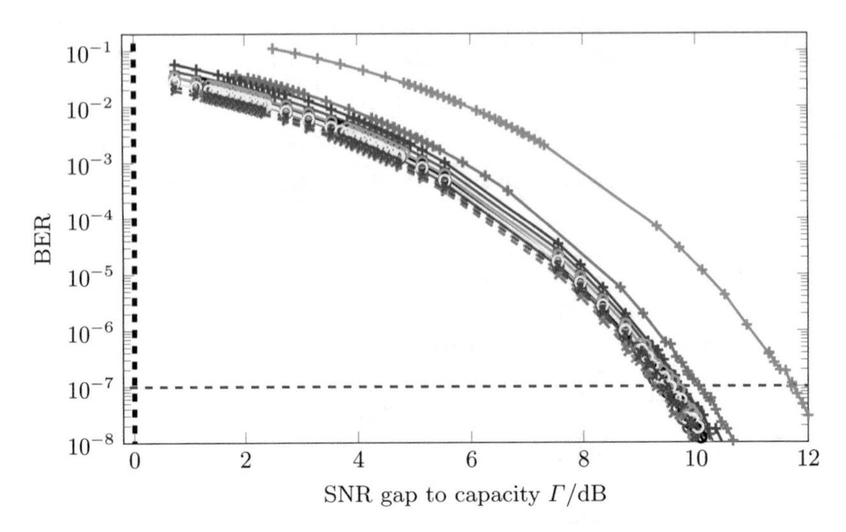

Fig. B.7 BER versus SNR plots for modulo receivers, uncoded QAM modulation, based on a hardware model of the G.fast receiver

B.3 SNR-Dependent BER for Trellis-Coded QAM Modulation

The SNR gap to capacity decreases when trellis coding, as described in Sect. 3.1.4, is used. However, trellis coding introduces some overhead, which is taken into account when calculating the gap to capacity. The effective number of bits transmitted with trellis-coded modulation is $b_{\text{eff}} = 0.75$ bits for the 1-bit constellation and $b_{\text{eff}} = \hat{b} - 0.5$ bits for the constellations with 2 or more bits. Thus, the SNR gap is calculated with respect to the number of modulated bits minus the trellis overhead, $b_{\text{eff}} = \log_2\left(1 + \frac{SNR}{\Gamma}\right)$.

Figures B.8 and B.9 show the SNR gap to capacity for trellis-coded modulation in comparison with uncoded modulation.

The SNR values to achieve the target bit error rate for each constellation are summarized in Table B.4 for linear receivers and in Table B.5 for modulo receivers.

B.4 SNR-Dependent BER for Trellis- and Reed–Solomon-Coding

With a combination of Reed–Solomon and trellis coding, even lower SNR is required to achieve the target bit error rate than for trellis-coding only. However, this comes with an increasing coding overhead, which is given by the ratio between the number of data bytes K_{rs} and the overall number of bytes per RS codeword N_{rs}. The effective

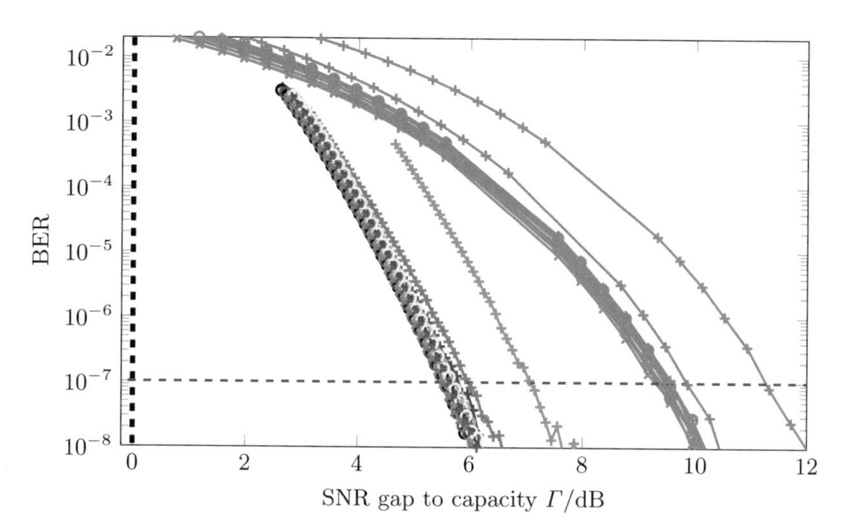

Fig. B.8 BER versus SNR curves for linear receivers, using trellis coded modulation, based on a hardware model of the G.fast receiver

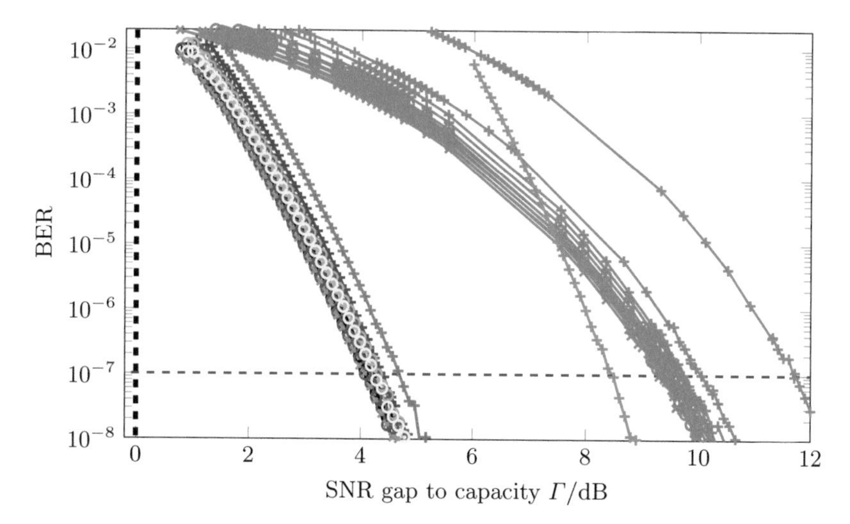

Fig. B.9 BER versus SNR plots for modulo receivers, using trellis coded modulation, based on a hardware model of the G.fast receiver

number of bits is $b_{\text{eff}} = 0.75\frac{K_{\text{rs}}}{N_{\text{rs}}}$ for the 1-bit constellation and $b_{\text{eff}} = \left(\hat{b} - 0.5\right)\frac{K_{\text{rs}}}{N_{\text{rs}}}$ for constellations with two or more bits.

Due to the additional overhead, the Reed–Solomon code does not necessarily reduce the gap to channel capacity. This is only the case when the Reed–Solomon overhead R_{rs} setting is selected correctly. Figures B.10 and B.11 show the SNR gap

Table B.4 SNR versus bit allocation table for trellis coded transmission, linear receiver

Bits	1	2	3	4	5	6	7
SNR	5.4	8.4	12.7	15.8	18.9	22.1	25.1
Bits	8	9	10	11	12	13	14
SNR	28.2	31.1	34.3	37.1	40.3	43.1	46.3

Table B.5 SNR versus bit allocation table for trellis coded transmission, modulo receiver

Bits	1	2	3	4	5	6	7
SNR	6.8	9.2	13.2	16.1	19.0	22.3	25.1
Bits	8	9	10	11	12	13	14
SNR	28.3	31.2	34.3	37.1	40.3	43.1	46.3

Table B.6 SNR versus bit allocation table for trellis+Reed Solomon transmission, linear receiver

Bits	1	2	3	4	5	6	7
SNR	2.8	5.6	10.0	13.3	16.4	19.7	22.6
Bits	8	9	10	11	12	13	14
SNR	25.8	28.7	31.9	34.8	37.9	40.8	43.9

Table B.7 SNR versus bit allocation table for trellis+Reed Solomon transmission, modulo receiver

Bits	1	2	3	4	5	6	7
SNR	4.8	6.8	10.8	13.8	16.6	19.9	22.7
Bits	8	9	10	11	12	13	14
SNR	25.9	28.8	31.9	34.8	37.9	40.8	43.9

to capacity for a 239/255 Reed–Solomon code with $R_{\text{rs}} = 16$ overhead bytes, which is the optimal selection for smaller constellations.

The required SNR values to achieve the target bit error rate of 10^{-7} are summarized in Tables B.6 and B.7. The difference between the SNR values of both tables is the modulo loss. Comparing the modulo loss over Appendices B.2, B.3 and B.4 shows that the modulo loss increases with increasing coding gain.

B.5 SNR Gap Approximation of the Discrete Rate Function

The capacity $b^{(k)}$ of carrier k on the additive white Gaussian noise channel is given by

$$b^{(k)} = \log_2\left(1 + SNR^{(k)}\right) \tag{B.15}$$

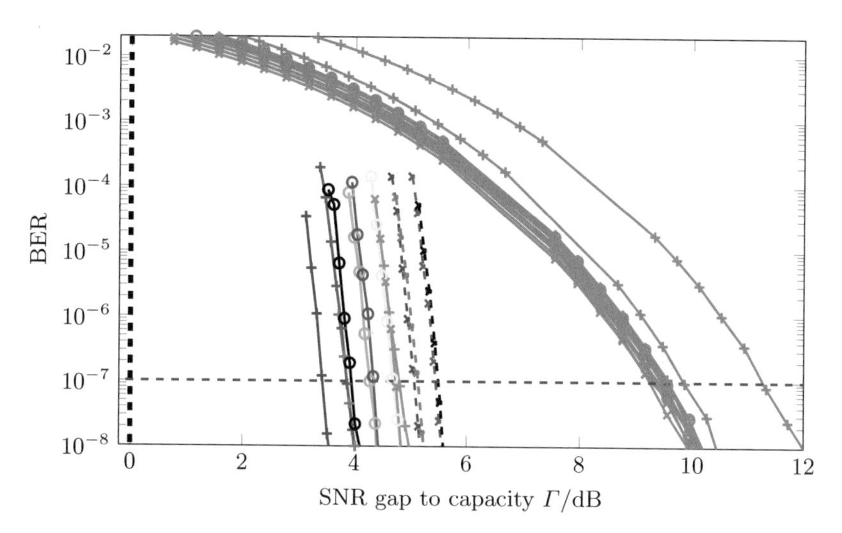

Fig. B.10 BER versus SNR plots for linear receivers, Reed–Solomon and trellis coding

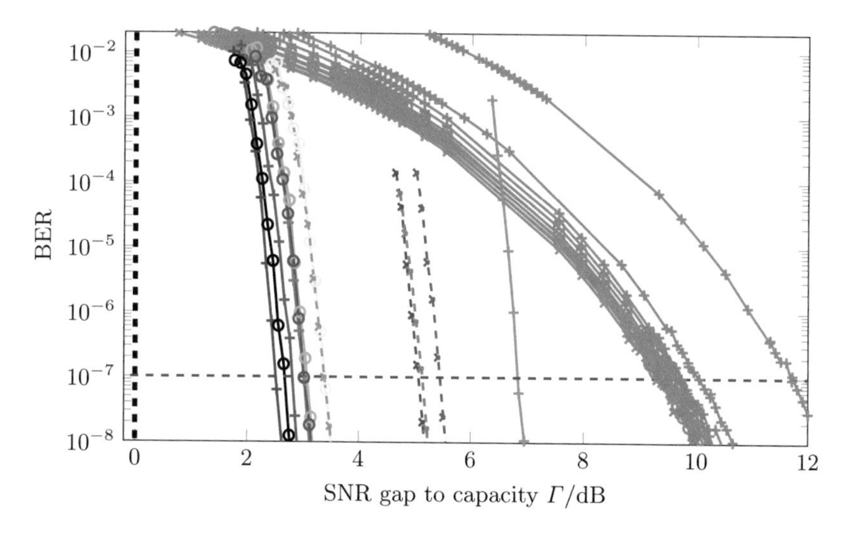

Fig. B.11 BER versus SNR plots for modulo receivers, Reed–Solomon and trellis coding

which is only achievable with Gaussian modulation. For systems with QAM modulation alphabet and imperfect channel coding, an approximation using the SNR gap Γ is used

$$b^{(k)} = \log_2\left(1 + \frac{SNR^{(k)}}{\Gamma}\right), \tag{B.16}$$

where the SNR gap γ accounts for imperfect coding with the coding gain γ_c and the SNR margin γ_m [2].

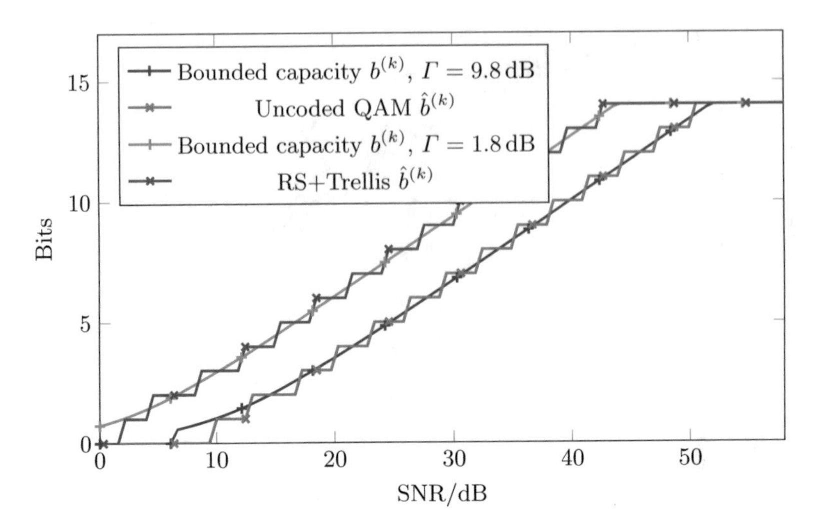

Fig. B.12 Comparison between bounded capacity with SNR gap and actual bit allocation for uncoded QAM modulation as well as for trellis and Reed–Solomon coding, ignoring the coding overhead

The actual bit allocation is a discrete function which is obtained by

$$\hat{b}^{(k)} = \max b \text{ s.t. } SNR_{\text{req}\,b} \leq \frac{SNR^{(k)}}{\gamma_{\text{m,min}}} \tag{B.17}$$

using the SNR versus constellation size tables, Tables B.2, B.3, B.4, B.5, B.6 and B.7.

The capacity $b^{(k)}$ according to Eq. (B.16) can be used as an upper bound for the bit allocation, but the optimization of the upper bound ignores the discrete nature of the actual rate objective. The additional SNR margin $\gamma_{\text{m}}^{(k)}$ on each carrier, which is a result of the integer bit allocation according to Eq. (B.17), can be translated into higher data rates, using the margin equalization techniques presented in Sect. 3.1.4.

With margin equalization, spectrum optimization with respect to the rate function according to Eq. (B.16) does not waste transmit power on carriers where the SNR margin is higher than the target margin. However, the margin equalization effect is not present for carriers where the SNR does not allow to transmit at least one bit, e.g., $b^{(k)} < 1$ and the excess margin in case that the maximum constellation size is exceeded, $b^{(k)} > b_{\max}$, cannot be used, either. Taking SNR lower bound $SNR_{\min}$ and the SNR upper bound $SNR_{\max}$ into account in the continuous rate function gives

$$b^{(k)} = \begin{cases} 0 & SNR^{(k)} \leq SNR_{\min} \\ \log_2\left(1 + \frac{SNR^{(k)}}{\Gamma}\right) & SNR_{\min} < SNR^{(k)} < SNR_{\max} \\ b_{\max} & SNR^{(k)} \geq SNR_{\max} \end{cases} . \tag{B.18}$$

The rate function (B.18) is used as an objective for convex sum-rate optimization The bound $SNR_{\min}$ and $SNR_{\max}$ are selected with respect to the minimum and maximum bit allocation. With the parameters $SNR_{\min}$, $SNR_{\max}$ and Γ within Eq. (B.18), implementation limitations are considered in the rate optimization.

The rate function according to Eq. (B.18) is concave within $SNR_{\min}$ and $SNR_{\max}$, which allows to use convex optimization algorithms for precoding and spectrum optimization, as presented in Sect. 3.2.8. These algorithms outperform greedy algorithms as in [3] and in contrast to the method described in [4], they are still implementable with reasonable effort.

B.6 Orthogonal Sequences for G.fast Channel Estimation

MIMO channel estimation in G.fast is performed by transmission of orthogonal sequences $w_l^{[t]}$ for line $l = 1, \ldots, L$ with a sequence length T, modulated on the sync symbols. Channel estimation is performed according to

$$\boldsymbol{H}_{\mathrm{est}}^{(k)} = \frac{1}{T} \sum_{t=1}^{T} \boldsymbol{e}^{(k)[t]} \boldsymbol{u}^{(k)[t],\mathrm{H}} + \boldsymbol{I}. \tag{B.19}$$

The orthogonal sequence elements have the values $w_l^{[t]} \in \{-1, 0, 1\}$. For the MIMO channel estimation as shown above, it is required that the sequences of all lines are mutually orthogonal

$$\sum_{t=1}^{T} w_l^{[t]} w_m^{[t]} = 0 \, \forall \, l \neq m. \tag{B.20}$$

The orthogonal sequence can be written as a matrix $\boldsymbol{W} \in \{-1, 0, 1\}^{L \times T}$. One way to create such matrix are Walsh–Hadamard matrices. They are created iteratively with the base matrix

$$\boldsymbol{W}_2 = \begin{bmatrix} 1 & 1 \\ 1 & -1 \end{bmatrix} \tag{B.21}$$

which extended iteratively to double size according to

$$\boldsymbol{W}_{2n} = \begin{bmatrix} \boldsymbol{W}_n & \boldsymbol{W}_n \\ \boldsymbol{W}_n & -\boldsymbol{W}_n \end{bmatrix}. \tag{B.22}$$

Walsh–Hadamard matrices consist of the values +1 and −1 and are square matrices of size $2^n \times 2^n$. An alternative way to construct orthogonal sequences is based on a pseudorandom sequence $w_{\text{pn}} \in \{-1, 1\}^{T-1}$ or a Legendre sequence [5] according to

$$\boldsymbol{W}_T = \begin{bmatrix} 1 & 1 & 1 & \dots & 1 \\ 1 & w_{\text{pn}}^{[1]} & w_{\text{pn}}^{[2]} & \dots & w_{\text{pn}}^{[T-1]} \\ 1 & w_{\text{pn}}^{[2]} & w_{\text{pn}}^{[3]} & \dots & w_{\text{pn}}^{[1]} \\ \dots & & & & \\ 1 & w_{\text{pn}}^{[T-1]} & w_{\text{pn}}^{[1]} & \dots & w_{\text{pn}}^{[T-2]} \end{bmatrix}. \tag{B.23}$$

These exist not only at the size of a 2^n, but also at a size of $4n$, created from a pseudorandom sequence of length $2^n - 1$ or a Legendre sequence of length $4n - 1$. Orthogonal sequences created from Legendre or pseudorandom sequences have advantages in terms of convergence of LMS algorithms for precoder adaptation.

The G.fast-standard allows to use 0-elements in the orthogonal sequence. The orthogonal base matrix is then

$$\boldsymbol{W}_2 = \begin{bmatrix} 1 & 0 \\ 0 & 1 \end{bmatrix}, \tag{B.24}$$

which can be extended according to

$$\boldsymbol{W}_{2n} = \begin{bmatrix} \boldsymbol{W}_n & \boldsymbol{0} \\ \boldsymbol{0}_n & \boldsymbol{W}_n \end{bmatrix}. \tag{B.25}$$

Sequences constructed according to (B.25) allow to perform channel estimation for independent groups of lines where one group is sending nonzero sync symbols while the other groups send zero symbols.

B.7 Line Joining for Zero-Forcing Precoding

For linear zero-forcing precoding in downstream, it is possible to perform line joining without a separate storage for the channel estimation. Starting with the initial precoder for the active lines $\boldsymbol{P}_{\text{aa}}^{(k),'}$, for which $\boldsymbol{G}_{\text{aa}}^{(k)} \boldsymbol{H}_{\text{aa}}^{(k)} \boldsymbol{P}_{\text{aa}}^{(k),'} = \boldsymbol{I}$ holds, the precoder matrix is extended to a larger matrix

$$\boldsymbol{P} = \begin{bmatrix} \boldsymbol{P}_{\text{aa}}^{(k)} & \boldsymbol{P}_{\text{aj}}^{(k)} \\ \boldsymbol{P}_{\text{ja}}^{(k)} & \boldsymbol{P}_{\text{jj}}^{(k)} \end{bmatrix} \tag{B.26}$$

in two steps. In the first step, the Vector-1 phase, the channel estimation is performed for the active lines as well as for the crosstalk from joining lines into active lines. The channel estimation for this case with active precoder gives

$$\boldsymbol{H}^{(k)}_{\text{est,v1}} \approx \boldsymbol{G}^{(k)}_{\text{aa}} \begin{bmatrix} \boldsymbol{H}^{(k)}_{\text{aa}} & \boldsymbol{H}^{(k)}_{\text{aj}} \end{bmatrix} \begin{bmatrix} \boldsymbol{P}^{(k),'}_{\text{aa}} & \boldsymbol{0} \\ \boldsymbol{0} & \boldsymbol{I} \end{bmatrix}. \tag{B.27}$$

After the Vector-1 phase, crosstalk between active lines as well as crosstalk from joining into active lines will be canceled according to

$$\boldsymbol{G}^{(k)}_{\text{aa}} \begin{bmatrix} \boldsymbol{H}^{(k)}_{\text{aa}} & \boldsymbol{H}^{(k)}_{\text{aj}} \end{bmatrix} \begin{bmatrix} \boldsymbol{P}^{(k),'}_{\text{aa}} & \boldsymbol{P}^{(k)}_{\text{aj}} \\ \boldsymbol{0} & \boldsymbol{I} \end{bmatrix} = \boldsymbol{I}. \tag{B.28}$$

The precoder coefficients to cancel crosstalk from joining into active lines are given by

$$\boldsymbol{P}^{(k)}_{\text{aj}} = -\boldsymbol{P}^{(k),'}_{\text{aa}} \boldsymbol{H}^{(k)}_{\text{est,v1,aj}} \tag{B.29}$$

while the precoder coefficients for crosstalk between the active lines $\boldsymbol{P}^{(k),'}_{\text{aa}}$ remain unchanged. With the precoder according to Eq. (B.29), it is possible to transmit data from the DP to the joining lines CPEs without disturbing the active lines. This is required to communicate initial settings to the joining CPEs and to setup the feedback channel which is required to train the full crosstalk canceler. Furthermore, the joining lines can initialize the receive equalizer $\boldsymbol{G}^{(k)}_{jj}$.

Then, the Vector-2 phase is performed where the vectoring feedback channel from the joining lines CPEs to the DPU is active and a full crosstalk channel estimation is performed. This gives the channel estimation matrices according to

$$\boldsymbol{H}^{(k)}_{\text{est,v2}} \approx \boldsymbol{G}^{(k)} \begin{bmatrix} \boldsymbol{H}^{(k)}_{\text{aa}} & \boldsymbol{H}^{(k)}_{\text{aj}} \\ \boldsymbol{H}^{(k)}_{\text{ja}} & \boldsymbol{H}^{(k)}_{\text{jj}} \end{bmatrix} \begin{bmatrix} \boldsymbol{P}^{(k),'}_{\text{aa}} & \boldsymbol{P}^{(k)}_{\text{aj,v1}} \\ \boldsymbol{0} & \boldsymbol{I} \end{bmatrix}. \tag{B.30}$$

The joining lines coefficients $\mathbf{P}^{(k)}_{\text{jj}}$ are given by

$$\mathbf{P}^{(k)}_{\text{jj}} = (\mathbf{H}^{(k)}_{\text{est,v2,jj}} - \mathbf{H}^{(k)}_{\text{est,v2,ja}} \mathbf{H}^{(k)}_{\text{est,v2,aj}})^{-1} \tag{B.31}$$

and the remaining precoder coefficients $\mathbf{P}^{(k)}_{\text{ja}}$ are given by

$$\mathbf{P}^{(k)}_{\text{ja}} = -\mathbf{P}^{(k)}_{\text{jj}} \mathbf{H}^{(k)}_{\text{est,v2,ja}}. \tag{B.32}$$

A correction of the active lines precoder coefficients $\boldsymbol{P}^{(k),'}_{\text{aa}}$ is required according to

$$\boldsymbol{P}^{(k)}_{\text{aa}} = \boldsymbol{P}^{(k),'}_{\text{aa}} (\boldsymbol{I} - \mathbf{H}^{(k)}_{\text{est,v2,aj}} \mathbf{P}^{(k)}_{\text{ja}}). \tag{B.33}$$

When the training is done according to this steps, line joining for linear zero-forcing precoding can be implemented without using additional memories.

B.8 Line Joining for Zero-Forcing Equalization

Based on the assumption of zero-forcing equalization, $\boldsymbol{G}^{(k)}\boldsymbol{H}^{(k)} = \boldsymbol{I}$, line joining algorithms for upstream can be derived similar to the downstream case in Sect. B.7. In contrast to the downstream direction, the complete upstream channel estimation $\boldsymbol{H}_{\text{est}}^{(k)}$ is available at the initial upstream training phase.

The matrix partitioning according to

$$\boldsymbol{G} = \begin{bmatrix} \boldsymbol{G}_{\text{aa}}^{(k)} & \boldsymbol{G}_{\text{aj}}^{(k)} \\ \boldsymbol{G}_{\text{ja}}^{(k)} & \boldsymbol{G}_{\text{jj}}^{(k)} \end{bmatrix} \tag{B.34}$$

is used for the full equalizer matrix, including active and joining lines. The equalizer matrix prior to joining is denoted as $\boldsymbol{G}_{\text{aa}}^{(k),\prime}$.

The following calculation steps are performed during the first upstream training phase R-VECTOR-1. A refinement can be done during the following upstream training phases. In the first step, the equalizer coefficients for crosstalk between joining lines are given by

$$\boldsymbol{G}_{\text{jj}}^{(k)} = (\boldsymbol{H}_{\text{est,jj}}^{(k)} - \boldsymbol{H}_{\text{est,ja}}^{(k)}\boldsymbol{G}_{\text{aa}}^{(k),\prime}\boldsymbol{H}_{\text{est,aj}}^{(k)})^{-1}. \tag{B.35}$$

The next step is to calculate equalizer coefficients from joining lines into active lines

$$\boldsymbol{G}_{\text{aj}}^{(k)} = -\boldsymbol{G}_{\text{aa}}^{(k),\prime}\boldsymbol{H}_{\text{est,aj}}^{(k)}\boldsymbol{G}_{\text{jj}}^{(k)}. \tag{B.36}$$

Then, the coefficients from active lines into joining lines are calculated

$$\boldsymbol{G}_{\text{ja}}^{(k)} = -\boldsymbol{G}_{\text{jj}}^{(k)}\boldsymbol{H}_{\text{est,ja}}^{(k)}\boldsymbol{G}_{\text{aa}}^{(k),\prime} \tag{B.37}$$

and finally, the coefficients between the active lines are corrected according to

$$\boldsymbol{G}_{\text{aa}}^{(k)} = \boldsymbol{G}_{\text{aa}}^{(k),\prime} - \boldsymbol{G}_{\text{aa}}^{(k),\prime}\boldsymbol{H}_{\text{est,aj}}^{(k)}\boldsymbol{G}_{\text{ja}}^{(k)}. \tag{B.38}$$

B.9 Algorithms for Spectrum Optimization with ZF Precoding

The spectrum optimization algorithm according to Sect. 3.2.8 gives an efficient way to solve the spectrum optimization problem optimally using Lagrange duality. The full algorithm is summarized in Algorithm 9.

The nested loops of updating the Lagrange multipliers of the sum-power constraint and performing spectrum optimization for given sum-power Lagrangian multipliers dominates the run-time of the algorithm. To overcome that, a 2-step optimization approach is proposed, which gives a solution that is very close to the optimum,

Algorithm 9 Spectrum optimization for the linear zero-forcing Precoder

Initialize $\boldsymbol{\mu}_{\text{sum}} = \mathbf{0}_L$
repeat
  for $k = 1$ **to** K **do**
    Start with all lines active $\mathbb{I}_{\text{a}} = \{1, \ldots, L\}$
    repeat
      Calculate ZF precoder $\boldsymbol{P}^{(k)}$ for given $\mathbb{I}_{\text{a}}$
      Initialize $\boldsymbol{x}^{(k)}$
      Calculate constraint set Eqs. (3.55) and (3.56)
      repeat
        Calculate quadratic approximation using (3.63) and (3.64) and solve (3.66)
      until Convergence of $\boldsymbol{x}^{(k)}$
      if $\exists SNR_l^{(k)} < SNR_{\min} : l \in \mathbb{I}_{\text{a}}^{(k)}$ **then**
        Remove the weakest link $\arg\min_{l \in \mathbb{I}_{\text{a}}^{(k)}} \left| \boldsymbol{h}_l^{(k),\text{T}} \boldsymbol{p}_l^{(k)} \right|^2$ from the set of active lines
      end if
    until All active carriers meet SNR lower bound
  end for
  Update power allocation using (3.67)
until Per-line sum-power is converged

but does not require this type of iterations. In the first step, the per-line spectrum optimization problem similar to Algorithm 9 in Sect. 3.2.8 is solved.

The first derivative for quadratic approximation of the rate objective without the sum-power constraint changes from (3.63) to

$$\frac{\partial \Phi(\boldsymbol{x}^{(k)})}{\partial x_l^{(k)}} = -\frac{1}{x_l^{(k)} + \Gamma \sigma^2 |g_{ll}^{(k)}|^2}, \tag{B.39}$$

while the second derivative is still given by Eq. (3.64).

This gives the first step of the 2-step spectrum optimization algorithm, which takes care of the per-line spectral mask constraints and bit loading upper bounds, according to Algorithm 10. The power allocation resulting from Algorithm 10, which is the optimal transmit power allocation considering only spectral masks and the bit loading upper bound, is used as an upper bound $\boldsymbol{x}_{\text{mask}}^{(k)}$ for the second step of the algorithm.

In the second step, only the per-line sum-power constraint is considered. The spectrum optimization problem for liner zero-forcing precoding with a per-line sum-power constraint, only, has a closed-form solution for a given Lagrange multiplier $\boldsymbol{\mu}_{\text{sum}}$. It is given by

$$x_{\text{sp},l}^{(k)} = (\boldsymbol{\mu}_{\text{sum}}^{\text{T}} \boldsymbol{a}_{\text{sum},l}^{(k)})^{-1} - |g_{ll}^{(k)}|^2 \Gamma \sigma^2 \tag{B.40}$$

$\boldsymbol{a}_{\text{sum},l}^{(k)}$ is the lth column of the sum-power constraint matrix $\boldsymbol{A}_{\text{sum}}^{(k)} = \boldsymbol{P}^{(k)} \odot \boldsymbol{P}^{(k),*}$.

The final power values $\boldsymbol{x}$ are obtained by a projection step

Algorithm 10 Step 1 of the 2-step spectrum optimization algorithm

for $k = 1$ **to** K **do**
  Start with all lines active $\mathbb{I}_\mathrm{a} = \{1, \ldots, L\}$
  repeat
    Calculate ZF precoder $\boldsymbol{P}^{(k)}$ for given $\mathbb{I}_\mathrm{a}$
    Initialize $\boldsymbol{x}^{(k)}$
    Calculate constraint set Eqs. (3.55) and (3.56)
    repeat
      Calculate quadratic approximation using (B.39) and (3.64) and solve (3.66)
    until Convergence of $\boldsymbol{x}^{(k)}$
    if $\exists SNR_l^{(k)} < SNR_\mathrm{min} : l \in \mathbb{I}_\mathrm{a}^{(k)}$ **then**
      Remove the weakest link $\arg\min_{l\in\mathbb{I}_\mathrm{a}} \left|\boldsymbol{h}_l^{(k),\mathrm{T}} \boldsymbol{p}_l^{(k)}\right|^2$ from the set of active lines
    end if
  until All active carriers meet SNR lower bound
end for

$$x_l^{(k)} = \begin{cases} 0 & \text{for } x_{\mathrm{sp},l}^{(k)} \le 0 \\ x_{\mathrm{mask},l}^{(k)} & \text{for } x_{\mathrm{sp},l}^{(k)} \ge x_{\mathrm{mask},l}^{(k)} \\ x_{\mathrm{sp},l}^{(k)} & \text{otherwise} \end{cases} \tag{B.41}$$

which guarantees, that the solution satisfies not only the per-line sum-power constraint, but also the per-carrier constraint set according to Eq. (3.56). The Lagrange multiplier $\boldsymbol{\mu}_\mathrm{sum}$ is still calculated by the use of the gradient method according to Eq. (3.67).

Algorithm 11 Step 2 of the 2-step spectrum optimization algorithm

Initialize $\boldsymbol{\mu}_\mathrm{sum} = \boldsymbol{0}_L$
repeat
  for $k = 1$ **to** K **do**
    Evaluate Eqs. (B.40) and (B.41)
  end for
  Update power allocation using (3.67)
until Per-line sum-power is converged

The second step of the spectrum optimization algorithm is summarized in Algorithm 11. The complexity advantage of this method over Algorithm 9 results from the fact that the per-carrier optimization and the sum-power constraint optimization are solved only once, one after the other. In Algorithm 9, the per-carrier problem is solved multiple times in nested loops.

The sub-optimality of Algorithms 10 and 11 occurs when the upper bound $\boldsymbol{x}_\mathrm{mask}^{(k)}$, which is used for the projection step, is too strict.

B.10 Derivation of Precoder Gradient Updates

This appendix shows the LMS update steps required to adapt the zero-forcing linear precoder to small channel changes during the tracking mode. The error vector $\boldsymbol{e}^{(k)}$ for carrier k is given by

$$\begin{aligned}\boldsymbol{e}^{(k)} = \hat{\boldsymbol{u}}^{(k)} - \boldsymbol{u}^{(k)} &= \boldsymbol{G}^{(k)}\left(\boldsymbol{H}^{(k)}\boldsymbol{P}^{(k)}\boldsymbol{u}^{(k)} + \boldsymbol{n}^{(k)}\right) - \boldsymbol{u}^{(k)} \\ &= \boldsymbol{G}^{(k)}\boldsymbol{H}^{(k)}\boldsymbol{P}^{(k)}\boldsymbol{u}^{(k)} + \boldsymbol{G}^{(k)}\boldsymbol{n}^{(k)} - \boldsymbol{u}^{(k)}. \end{aligned} \tag{B.42}$$

The objective for the unconstrained MMSE precoder is given by

$$\begin{aligned}\mathrm{tr}\left(\boldsymbol{e}^{(k)}\boldsymbol{e}^{(k),\mathrm{H}}\right) = \mathrm{tr}\big(&\boldsymbol{G}^{(k)}\boldsymbol{H}^{(k)}\boldsymbol{P}^{(k)}\boldsymbol{u}^{(k)}\boldsymbol{u}^{(k),\mathrm{H}}\boldsymbol{G}^{(k),\mathrm{H}}\boldsymbol{P}^{(k),\mathrm{H}}\boldsymbol{H}^{(k),\mathrm{H}} \\ &+ \boldsymbol{G}^{(k)}\boldsymbol{H}^{(k)}\boldsymbol{P}^{(k)}\boldsymbol{u}^{(k)}\boldsymbol{n}^{(k),\mathrm{H}}\boldsymbol{G}^{(k),\mathrm{H}} \\ &- \boldsymbol{G}^{(k)}\boldsymbol{H}^{(k)}\boldsymbol{P}^{(k)}\boldsymbol{u}^{(k)}\boldsymbol{u}^{(k),\mathrm{H}} \\ &+ \boldsymbol{G}^{(k)}\boldsymbol{n}^{(k)}\boldsymbol{u}^{(k),\mathrm{H}}\boldsymbol{G}^{(k),\mathrm{H}}\boldsymbol{P}^{(k),\mathrm{H}}\boldsymbol{H}^{(k),\mathrm{H}} \\ &+ \boldsymbol{G}^{(k)}\boldsymbol{n}^{(k)}\boldsymbol{n}^{(k),\mathrm{H}}\boldsymbol{G}^{(k),\mathrm{H}} - \boldsymbol{G}^{(k)}\boldsymbol{n}^{(k)}\boldsymbol{u}^{(k),\mathrm{H}} \\ &- \boldsymbol{u}^{(k)}\boldsymbol{u}^{(k),\mathrm{H}}\boldsymbol{G}^{(k),\mathrm{H}}\boldsymbol{P}^{(k),\mathrm{H}}\boldsymbol{H}^{(k),\mathrm{H}} \\ &- \boldsymbol{u}^{(k)}\boldsymbol{n}^{(k),\mathrm{H}}\boldsymbol{G}^{(k),\mathrm{H}} + \boldsymbol{u}^{(k)}\boldsymbol{u}^{(k),\mathrm{H}}\big). \end{aligned} \tag{B.43}$$

This gives the LMS update step to be

$$\begin{aligned}\frac{\partial \mathrm{tr}\left(\boldsymbol{e}^{(k)}\boldsymbol{e}^{(k),\mathrm{H}}\right)}{\partial \boldsymbol{P}^{(k),*}} &= \boldsymbol{H}^{(k),\mathrm{H}}\boldsymbol{G}^{(k),\mathrm{H}}\boldsymbol{G}^{(k)}\boldsymbol{H}^{(k)}\boldsymbol{P}^{(k)}\boldsymbol{u}^{(k)}\boldsymbol{u}^{(k),\mathrm{H}} \\ &+ \boldsymbol{H}^{(k),\mathrm{H}}\boldsymbol{G}^{(k),\mathrm{H}}\boldsymbol{G}^{(k)}\boldsymbol{n}^{(k)}\boldsymbol{u}^{(k),\mathrm{H}} + \boldsymbol{H}^{(k),\mathrm{H}}\boldsymbol{G}^{(k),\mathrm{H}}\boldsymbol{u}^{(k)}\boldsymbol{u}^{(k),\mathrm{H}} \\ &= \boldsymbol{H}^{(k),\mathrm{H}}\boldsymbol{G}^{(k),\mathrm{H}}\boldsymbol{e}^{(k)}\boldsymbol{u}^{(k),\mathrm{H}}. \end{aligned} \tag{B.44}$$

The steady state solution of an LMS algorithm is

$$\begin{aligned} \mathrm{E}\left[\boldsymbol{H}^{(k),\mathrm{H}}\boldsymbol{G}^{(k),\mathrm{H}}\boldsymbol{e}^{(k)}\boldsymbol{u}^{(k),\mathrm{H}}\right] &= \boldsymbol{0} \\ \Rightarrow \boldsymbol{G}^{(k)}\boldsymbol{H}^{(k)}\boldsymbol{P}^{(k)}\mathrm{E}\left[\boldsymbol{u}^{(k)}\boldsymbol{u}^{(k),\mathrm{H}}\right] + \mathrm{E}\left[\boldsymbol{n}^{(k)}\boldsymbol{u}^{(k),\mathrm{H}}\right] &= \mathrm{E}\left[\boldsymbol{u}^{(k)}\boldsymbol{u}^{(k),\mathrm{H}}\right] \\ \Rightarrow \boldsymbol{P}^{(k)} &= \left(\boldsymbol{G}^{(k)}\boldsymbol{H}^{(k)}\right)^{-1}, \end{aligned} \tag{B.45}$$

which is the inverse of the channel. To keep the precoder coefficients within a predefined dynamic range, the normalization of Eq. (3.94) can be used or the diagonal elements of the precoder are excluded from the LMS update.

An alternative precoder update step, which does not require the channel matrix $\boldsymbol{H}^{(k)}$ is given by

$$\boldsymbol{P}^{(k),[t+1]} = \boldsymbol{P}^{(k),[t]} - \alpha_P \boldsymbol{e}^{(k),[t]}\boldsymbol{u}^{(k),[t]}. \tag{B.46}$$

For square precoder and channel matrices, it converges to the same steady state solution

$$\begin{aligned} \mathrm{E}\left[\boldsymbol{e}^{(k)}\boldsymbol{u}^{(k),\mathrm{H}}\right] &= \mathbf{0} \\ \Rightarrow \boldsymbol{G}^{(k)}\boldsymbol{H}^{(k)}\boldsymbol{P}^{(k)}\mathrm{E}\left[\boldsymbol{u}^{(k)}\boldsymbol{u}^{(k),\mathrm{H}}\right] + \mathrm{E}\left[\boldsymbol{n}^{(k)}\boldsymbol{u}^{(k),\mathrm{H}}\right] &= \mathrm{E}\left[\boldsymbol{u}^{(k)}\boldsymbol{u}^{(k),\mathrm{H}}\right] \\ \Rightarrow \boldsymbol{P}^{(k)} &= \left(\boldsymbol{G}^{(k)}\boldsymbol{H}^{(k)}\right)^{-1}. \end{aligned} \tag{B.47}$$

B.11 Gradient Spectrum Update for Nonlinear ZF Precoding

For zero-forcing nonlinear precoding, the gradient-based spectrum optimization may be used in the tracking phase as an alternative to Algorithm 6 in Sect. 3.3.3. The main advantage is the more accurate estimation of the power loss. The algorithm is summarized in Algorithm 12.

Algorithm 12 Nonlinear zero-forcing precoding gradient update

---

Initialize $\boldsymbol{\mu}_{\mathrm{sum}} = \mathbf{0}_L$
repeat
  for $k = 1$ **to** K **do**
    Receive error vector $\boldsymbol{e}^{(k)}$ from CPE side
    Estimate power loss $\boldsymbol{P}_{\mathrm{m}}^{(k)}$ using Eq. (3.109)
    Update spectrum (Eq. (3.95))
    Update constraint set Eq. (3.119)
    repeat
      Evaluate constraint set violations $\mathbb{I}_{\mathrm{violated}}$
      Perform projection Eq. (3.96)
    until Constrains are satisfied
  end for
  Update power allocation using (3.67)
until Per-line sum-power is converged

B.12 Zero-Forcing Approximation of Bit Loading Constraint

Linear MMSE precoding as discussed in Sects. 3.2.9 and 3.2.10 use a zero-forcing approximation of the bit loading upper bound,

$$\frac{x_l^{(k)}\left|\boldsymbol{h}_l^{(k),\mathrm{T}}\boldsymbol{p}_l^{(k)}\right|^2}{\sigma^2} \leq \Gamma(2^{b_{\max}} - 1). \tag{B.48}$$

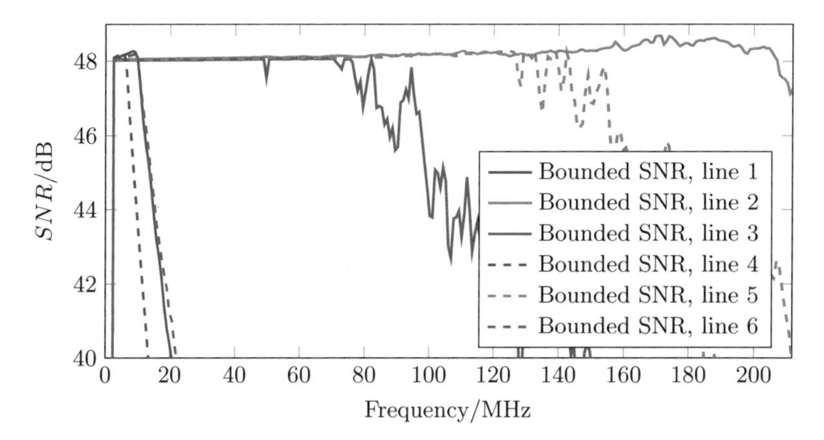

Fig. B.13 Actual SNR for MMSE precoding with zero-forcing constraint approximation of the maximum SNR constraint at $SNR_{\max} = 48\,\text{dB}$

The same approximation is used for G.fast channel capacity in Sect. 3.5. The approximation results in a convex power constraint, which is more strict than the original SNR upper bound, which would be

$$\frac{x_l^{(k)}\left|\boldsymbol{h}_l^{(k),\mathrm{T}}\boldsymbol{p}_l^{(k)}\right|^2}{\left(\sigma^2+\sum\limits_{d>l}x_d^{(k)}\left|\boldsymbol{h}_l^{(k),\mathrm{T}}\boldsymbol{p}_d^{(k)}\right|^2\right)} \le \Gamma(2^{b_{\max}}-1) \tag{B.49}$$

for nonlinear precoding and

$$\frac{x_l^{(k)}\left|\boldsymbol{h}_l^{(k),\mathrm{T}}\boldsymbol{p}_l^{(k)}\right|^2}{\left(\sigma^2+\sum\limits_{d\neq l}x_d^{(k)}\left|\boldsymbol{h}_l^{(k),\mathrm{T}}\boldsymbol{p}_d^{(k)}\right|^2\right)} \le \Gamma(2^{b_{\max}}-1) \tag{B.50}$$

for linear precoding. The approximation is based on the condition

- $\sum\limits_{d\neq l} x_d|\boldsymbol{h}_l^{(k),\mathrm{T}}\boldsymbol{p}_d^{(k)}|^2 \ll \sigma^2$ for $SNR_l^{(k)} \geq \Gamma(2^{b_{\max}}-1)$ for linear precoding
- and $\sum\limits_{d>l} x_d|\boldsymbol{h}_l^{(k),\mathrm{T}}\boldsymbol{p}_d^{(k)}|^2 \ll \sigma^2$ for $SNR_l^{(k)} \geq \Gamma(2^{b_{\max}}-1)$ for nonlinear precoding.

This holds true when the precoder converges to the zero-forcing solution in the high SNR region.

In case that the zero-forcing approximation does not hold, the actual SNR would be below the target SNR. Observing the SNR as it appears when performing MMSE precoder optimization with the maximum SNR constraint, as shown in Fig. B.13, the SNR is not below the target SNR.

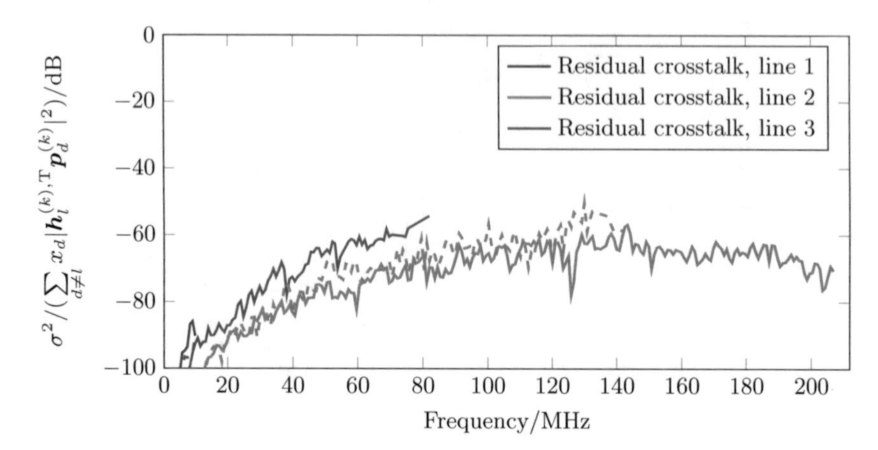

Fig. B.14 Ratio between residual crosstalk level and background noise σ^2

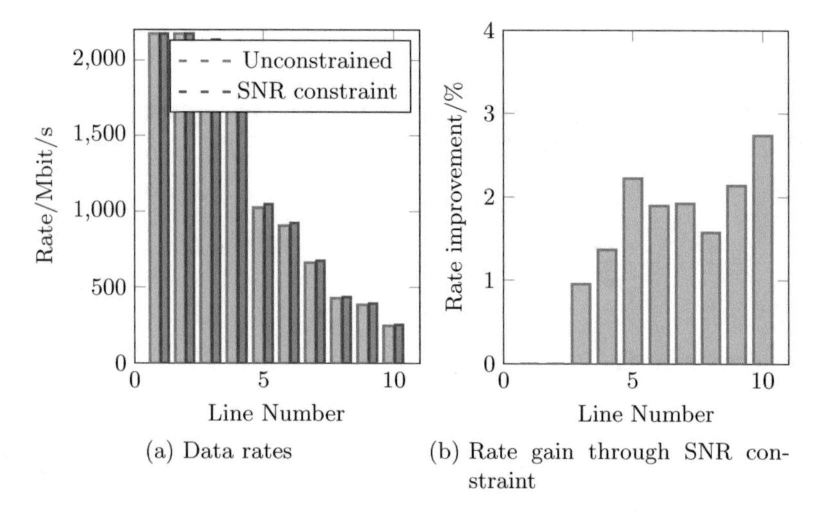

Fig. B.15 Comparison between weighted rate optimization and peak rates

Alternatively, the condition $\sum_{d\neq l} x_d|\boldsymbol{h}_l^{(k),\mathrm{T}}\boldsymbol{p}_d^{(k)}|^2 \ll \sigma^2$ can be checked, directly. The residual crosstalk level, normalized to the noise level σ^2 is shown in Fig. B.14.

Finally, the benefit of such constraint in terms of data rates is presented in Fig. B.15. The resulting data rates for this example show that the data rates increase slightly through the bit loading constraint, compared to MMSE precoding without the SNR constraint.

These examples show that the zero-forcing approximation of the maximum SNR constraint, as it is used in Sects. 3.2.9 and 3.5 does not cause any issue for the SNR regions of interest, which are around 45 dB for the standard G.fast transmission settings and around 35 dB for channel capacity, bounded to 12 bit.

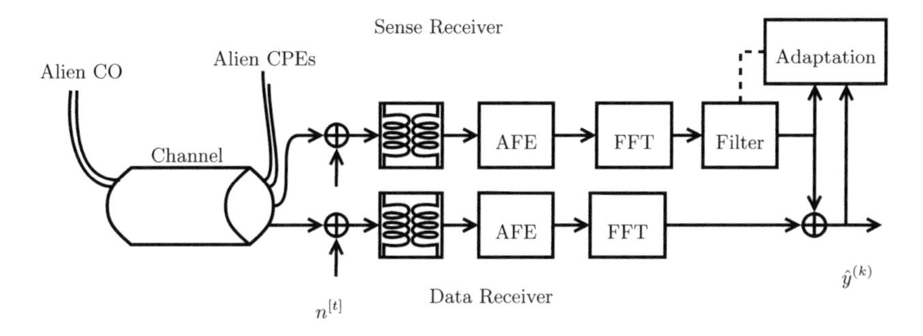

Fig. B.16 Receiver with alien crosstalk cancelation

B.13 Structure of Frequency Domain Alien Crosstalk Cancelation

Alien crosstalk cancelation in frequency domain can be implemented with lower compute complexity than the time domain approach according to Eq. (3.157) in Sect. 3.6.3. However, the frequency domain cancelation is not capable to cancel intercarrier interference. The block diagram of a G.fast receiver using frequency domain alien crosstalk cancelation is shown in Fig. B.16.

The alien crosstalk filter $\boldsymbol{h}_{\text{filter}} \in \mathbb{C}^K$ is applied according to

$$\hat{y}^{(k)}_{\text{out}} = \hat{y}^{(k)} - \hat{y}^{(k)}_{\text{sense}} h^{(k)}_{\text{filter}}. \tag{B.51}$$

and the coefficient update rule is given by

$$h^{(k),[t+1]}_{\text{filter}} = h^{(k),[t]}_{\text{filter},i} - \alpha_{\text{filter}} e^{(k),[t]} y^{(k),[t]}_{\text{sense}}. \tag{B.52}$$

B.14 Simulation Results for Alien Crosstalk Cancelation

The following simulation results show the performance of alien crosstalk cancelation in a scenario of short line length between 50 and 150 m of the DTAG-PE05 cable and co-located CPEs.

Comparing the results from Figs. B.17, B.18 and B.19 with co-located CPEs with the results for G.fast/VDSL2 coexistence with distributed CPEs in Appendix B.16 shows that the performance penalty is stronger in the case of co-located CPEs, which is a result of the stronger CPE NEXT. Alien crosstalk cancelation shows some performance benefits for this special case. However, the rate impact due to alien crosstalk is more severe for this setup than in the more practical network topology used in Appendix B.16.

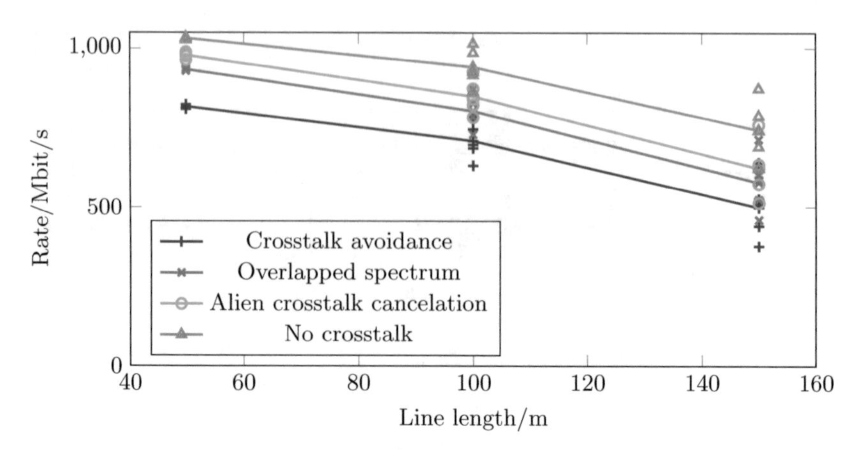

Fig. B.17 Rate versus reach curves for alien crosstalk cancelation on a DTAG-PE05 binder with co-located CPEs and one alien disturber

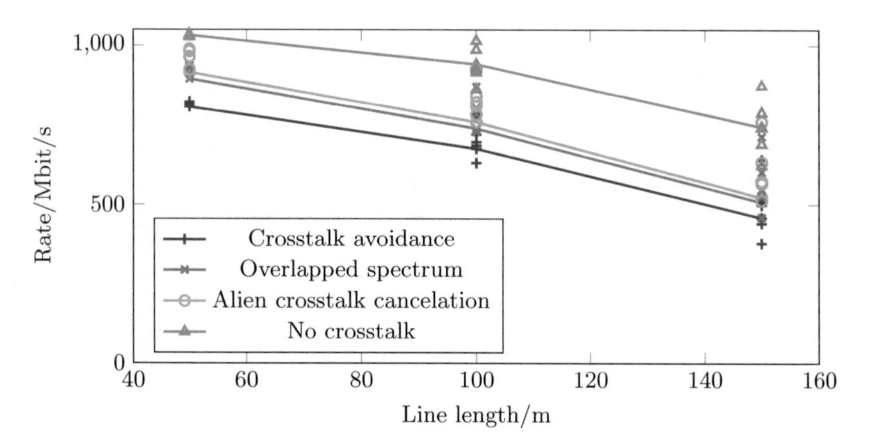

Fig. B.18 Rate versus reach curves for alien crosstalk cancelation on a DTAG-PE05 binder with co-located CPEs and two alien disturbers

As the lines at the DP side as well as at the CPE side are co-located and the lines are short, the crosstalk impact on downstream and upstream direction is almost equal and Figs. B.17, B.18 and B.19 show aggregate rates instead of separating downstream and upstream data rates.

The background noise is -150 dBm/Hz, the transmit power is 4 dBm for G.fast and 14.5 dBm for VDSL2, where G.fast uses the 106 MHz profile while VDSL2 uses the 17 MHz profile.

A 30-pair binder of the quad-structured DTAG-PE05 cable is used while two pairs of the quad are used for the G.fast data and sense line. The alien crosstalk lines are placed on the neighboring lines in the binder with strongest crosstalk coupling into the G.fast victim line.

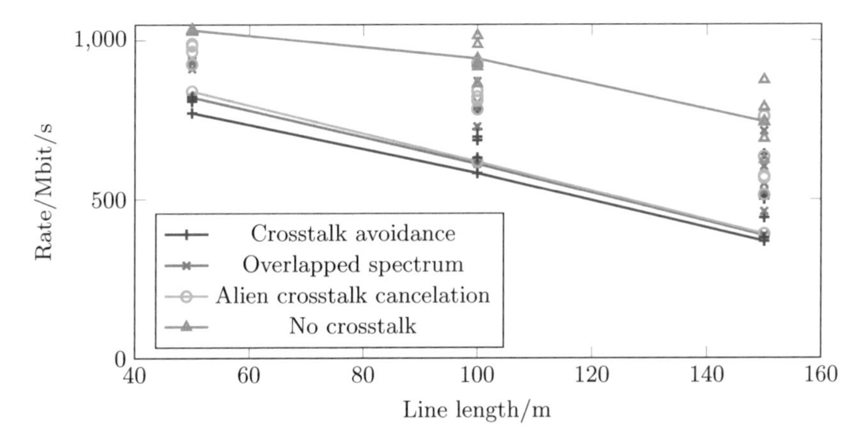

Fig. B.19 Rate versus reach curves for alien crosstalk cancelation on a DTAG PE05 binder with co-located CPEs and 5 alien disturbers

Observing the case of a single alien disturber in Fig. B.17, there is some gain of the overlapped spectrum with alien crosstalk cancelation over the case of plain overlapped spectrum. It is not possible for alien crosstalk cancelation to fully recover the performance loss due to alien crosstalk due to the fact that the sense line experiences some noise which is amplified through the canceler filter.

Increasing the number of alien disturbers from one to two in Fig. B.18 and to 5 disturbers in Fig. B.19 gives a situation where alien crosstalk cancelation gives almost no benefit over overlapped spectrum without alien crosstalk cancelation.

B.15 A View on Rate Regions and Rate-Versus-Reach Curves

Performance comparisons as well as the interpretation of rate regions are done differently in wireline communication than in wireless communication. One reason is a different technology evolution, legacy wireless technologies use TDMA or FDMA methods to access the channel, while in legacy wireline technologies subscribers use the cable binder like a non-shared medium, accepting some crosstalk between the subscribers. Another reason is that the channel quality of wireline subscribers depends mainly on the line length between DPU and CPE, which does not change over time.

The first wireline broadband transmission technologies such as SDSL and ADSL used low frequencies and did not experience much interference between the channels. ADSL and SDSL subscribers have parallel access to the transmission medium and there is only a small performance penalty to the individual subscriber due to the multi-user access. The rate region, assuming a user group with rates R_1 and a second

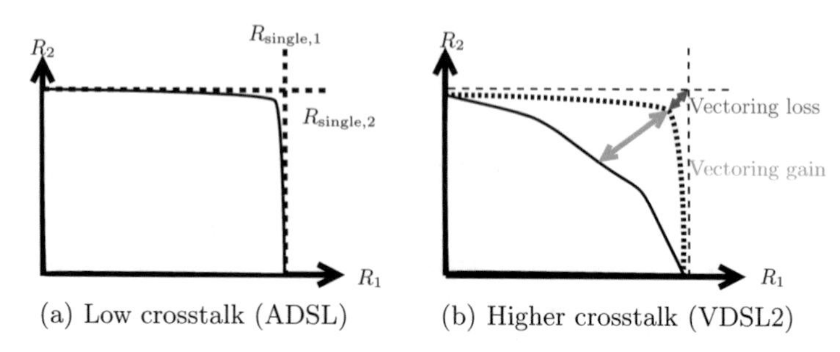

Fig. B.20 Rate regions with low crosstalk

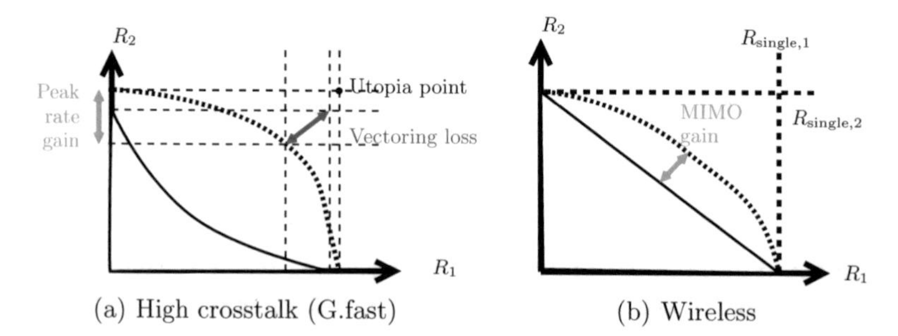

Fig. B.21 Rate regions with high crosstalk

group with rates R_2, is shaped as in Fig. B.20a and the achievable rates are close to the single line rates $R_{\text{single},1}$ and $R_{\text{single},2}$ of individual subscribers, which are used as a performance reference for spectrum management methods.

With increasing crosstalk, the rate loss increases and in crosstalk dominated DSL technologies such as VDSL2, the shape of the rate region changes as in Fig. B.20b. Crosstalk cancelation, called Vectoring in VDSL2, is introduced to increase performance. This gives a significant rate gain over the crosstalk limited case. However, the achievable single line data rates are still used as reference and crosstalk cancelation performance is measured by the gap between the Vectored rates and the single line rates, marked as "Vectoring loss".

Further increasing the frequencies in G.fast gives the rate region as shown in Fig. B.21a, where even crosstalk cancelation results in a substantial gap to the single line rates. However, due to high crosstalk couplings, transmit power can be focused on a subset of subscribers to increase their rate, even beyond the single line rates, as it is done in peak rate optimization in Sect. 4.4. The sum of all peak rates gives the Utopia point, which is beyond the single line rates point.

In wireless systems, on the other hand, orthogonal channel access is traditionally used, which gives the rate region as in Fig. B.21b. Multi-user MIMO systems extend the rate region, which gives the view of a gain through the MIMO processing.

In wireless, rate distributions and average data rates are a usual performance measure to compare different signal processing methods. Figure B.22b gives an example

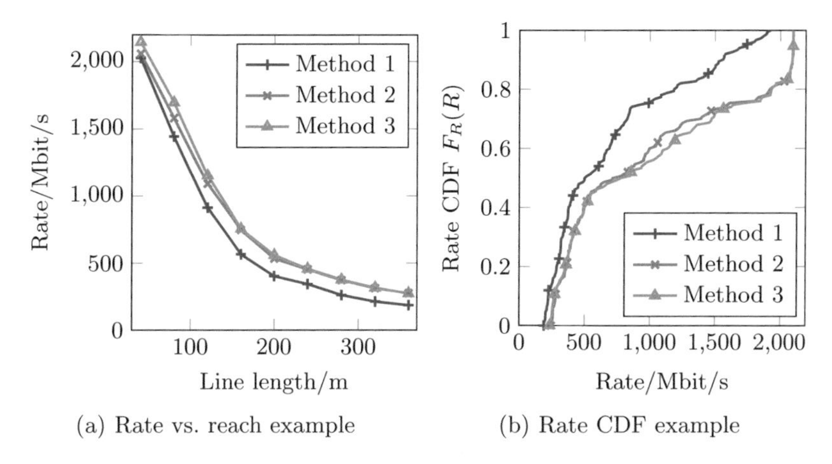

Fig. B.22 Rate versus reach curves and rate CDF for performance comparison between different algorithms

of rate cumulative densities for some example simulation on G.fast data rates. The reason is that the users move and therefore, experience different channel conditions over time.

In wireline, on the other hand, users don't move. Therefore, the rate vs. reach curve, showing average or minimum rates as a function of line length, is a widely used performance measure. Figure B.22a gives an example of the rate vs. reach curves corresponding to Fig. B.22b. In Appendices B.17–B.20, the results are reported in terms of rate versus reach curves.

B.16 Simulation Results for G.fast with VDSL Coexistence

The performance of G.fast operating with coexisting VDSL2 depends on the network topology, whether CPEs are co-located or non co-located and whether the VDSL cabinet and the G.fast DP are co-located or spatially separated. Different deployment scenarios are analyzed, here. Some of the results have been presented in [6].

The background noise is assumed to be −140 dBm/Hz below 30 MHz and −150 dBm/Hz above 30 MHz, the transmit power is 4 dBm for G.fast and 14.5 dBm for VDSL 2, where G.fast uses the 106 MHz profile while VDSL2 uses the 17 MHz profile and the DTAG-PE05 line according to Chap. 2 is used as a channel model.

Alien crosstalk impacts the downstream and the upstream direction differently. Therefore, the following plots show the G.fast downstream and upstream rates independently, while the ratio between downstream and upstream time is 0.7/0.3. Solid lines show the crosstalk avoidance case where the G.fast start frequency is 23 MHz while the dashed lines show the overlapped spectrum strategy without alien crosstalk cancelation, using a G.fast start frequency of 2 MHz according to [7].

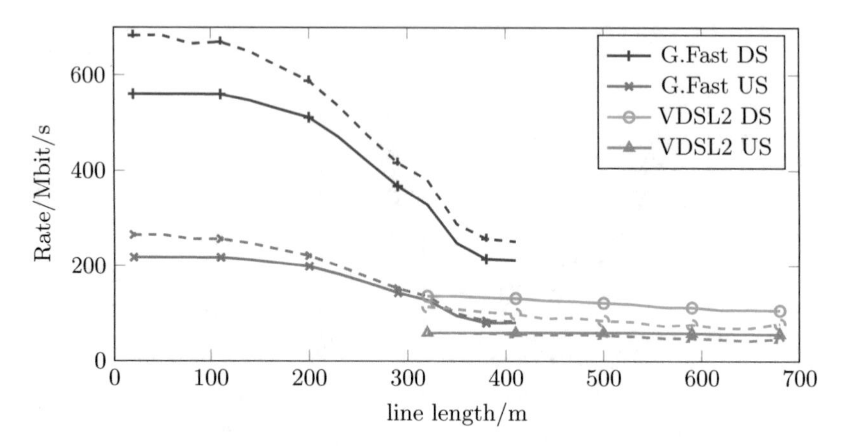

Fig. B.23 Rate versus reach curves for G.fast from the DP, coexisting with VDSL2 from the street cabinet, without power back-off, using the crosstalk avoidance strategy (solid lines) and overlapped spectrum (dashed lines)

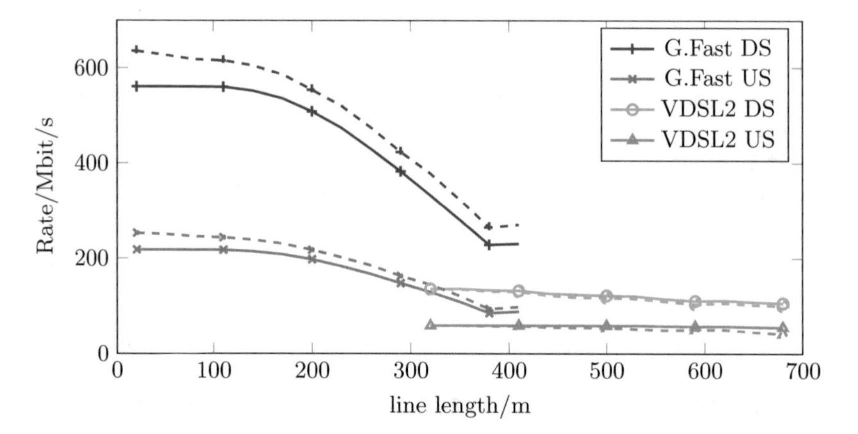

Fig. B.24 Rate versus reach curves for G.fast from the DP, coexisting with VDSL2 from the street cabinet, with power back-off on the G.fast lines to protect VDSL2, using the crosstalk avoidance strategy (solid lines) and overlapped spectrum (dashed lines)

The most relevant deployment scenario for G.fast with VDSL2 coexistence is G.fast from the distribution point coexisting with VDSL2 from the street cabinet as shown in Fig. 3.17a in Sect. 3.6. The results are shown in Fig. B.23 without VDSL protection and Fig. B.24 with VDSL2 protection by power back-off on the G.fast lines.

The rate versus reach curves in Figs. B.23 and B.24 range between 10 and 400 m for the G.fast service and between 400 and 700 m for the VDSL lines, indicating a 300 m line between VDSL cabinet and G.fast DP. This scenario gives substantial rate gains of the overlapped spectrum strategy (dashed lines) over the crosstalk avoidance case (solid lines), even with conservative power back-off settings as shown in

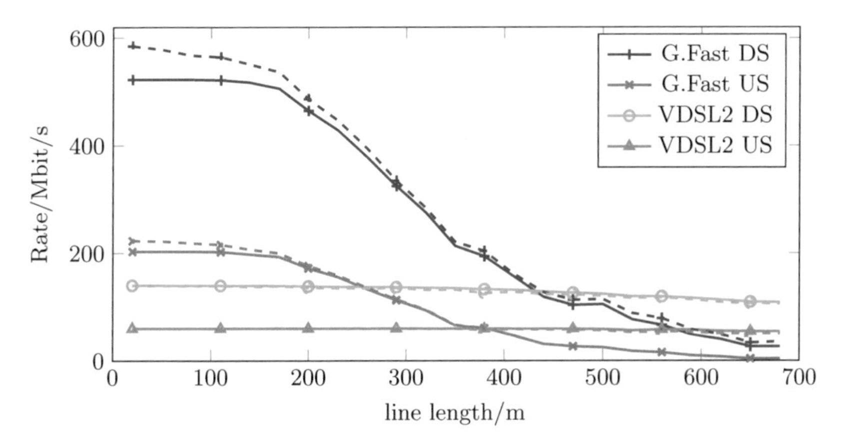

Fig. B.25 Long range G.fast and VDSL2 provided from the same FTTC location with G.fast power back-off to protect VDSL2, using the crosstalk avoidance strategy (solid lines) and overlapped spectrum (dashed lines)

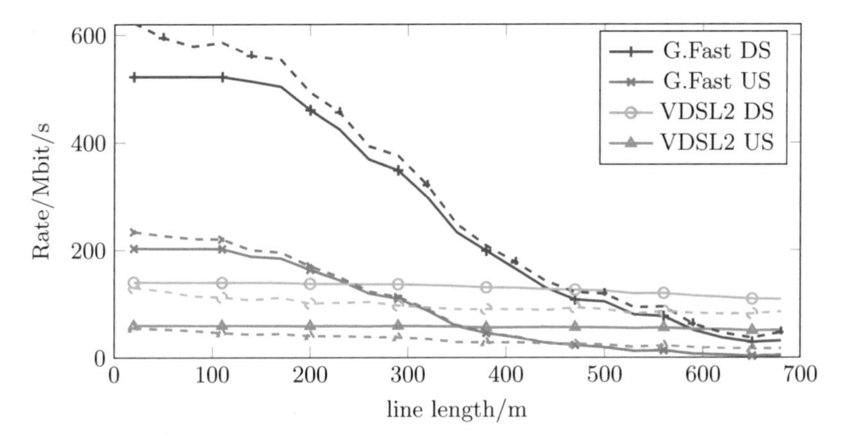

Fig. B.26 Long range G.fast and VDSL2 provided from the same FTTC location without power back-off, using the crosstalk avoidance strategy (solid lines) and overlapped spectrum (dashed lines)

Fig. B.24. Without power back-off, as in Fig. B.23, VDSL2 downstream data rates drop significantly due to the alien crosstalk.

An alternative deployment scenario with VDSL2 and G.fast provided from the same location, as shown in Fig. 3.17b, is evaluated in Figs. B.25 and B.26 where G.fast and VDSL are provided from the street cabinet and therefore, serve an increased line length up to 700 m. Special settings for G.fast are required in this case to support the extended range. The transmit power is increased from 4 to 8 dBm and the cyclic extension is increased from 320 samples to 640 samples to avoid intercarrier interference and intersymbol interference on longer loops.

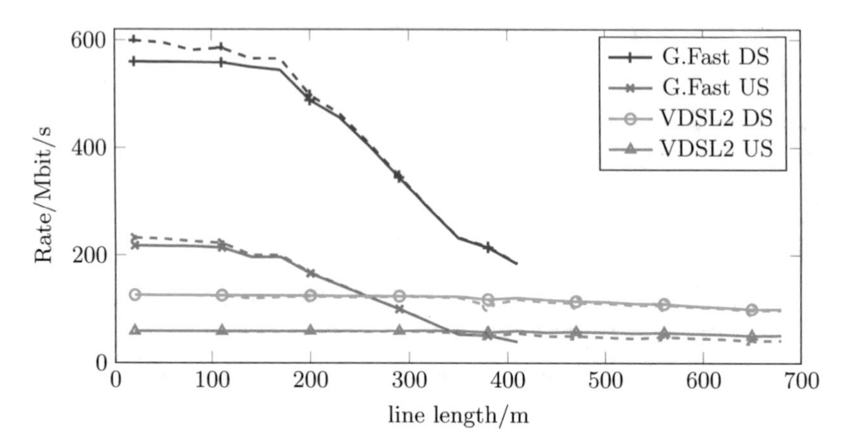

Fig. B.27 Standard G.fast and VDSL2 FTTC with power back-off, using the crosstalk avoidance strategy (solid lines) and overlapped spectrum (dashed lines)

Without VDSL2 protection, there is still some performance gain in the multi-mode cabinet scenario, as shown by Fig. B.26 while the data rates with G.fast power back-off drop below the VDSL rates at a line length around 400 m. In this case, there is no benefit from long range G.fast and the short range G.fast settings can be used, giving the rate vs. reach curves from Fig. B.27 where all lines with a length beyond 400 m are forced to use the VDSL2 service.

B.17 Simulations for Linear Precoding Methods

Section 3.2 presents various approaches to optimized linear precoding. The column norm scaling ([8] or Sect. 3.2.6) approach and the greedy rate optimization ([9] or Sect. 3.2.7) are low complexity methods to perform zero-forcing precoding and satisfy the spectral constraints.

The performance comparison is done in terms of rate versus reach curves. Two deployment scenarios are considered to create the rate versus reach curves, co-located CPEs as in Fig. 3.23a and non co-located CPEs as in Fig. 3.23b in Sect. 3.7. For co-located CPEs, the rate versus reach curve, e.g., Fig. B.28, is created from multiple simulation runs where each of them uses a cable binder with all lines at the same length. The rates achieved by individual lines are represented by the marks, which appear only at certain line length steps. For non co-located CPEs, e.g., in Fig. B.29, the simulation runs on multiple cable binders while each of the cable binders consists of lines with different line length. The line length is randomly distributed, as indicated by the marks appearing at any line length between the minimum and maximum.

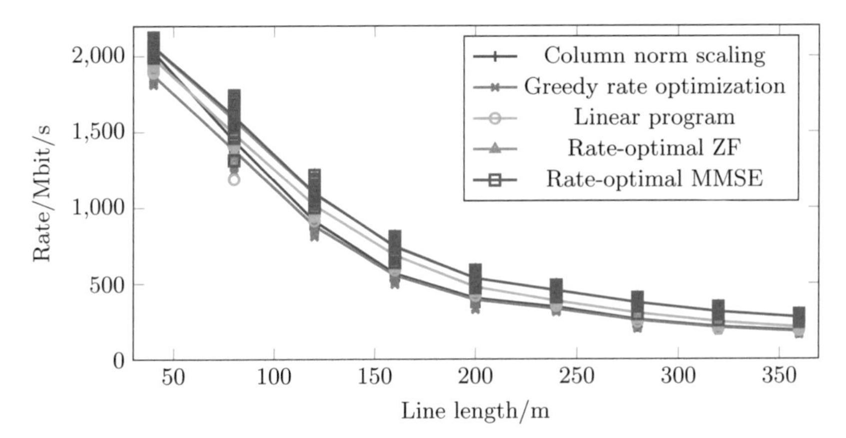

Fig. B.28 Rate versus reach curves for comparison of different linear precoding and spectrum optimization methods on a 30 pair DTAG-PE06 binder with co-located CPEs (Fig. 3.23a)

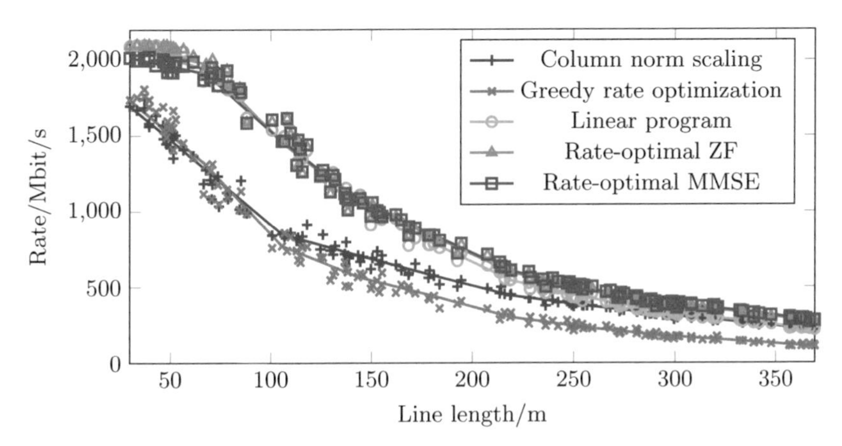

Fig. B.29 Rate versus reach curves for comparison different linear precoding and spectrum optimization methods on a 30 pair DTAG-PE06 binder with non-co-located CPEs (Fig. 3.23b)

Section 3.2.8 discusses spectrum optimization for zero-forcing precoding, based on convex optimization. Section 3.2.9 presents weighted MMSE precoding as an upper bound for the achievable performance of linear precoding methods.

The simulation conditions are a background noise of −140 dBm/Hz, 4 dBm transmit power, and a 212 MHz G.fast profile according to [7] with 2 MHz start frequency. Reed–Solomon and trellis coding with 6 dB target margin is applied, using a discrete bit loading between 1 bit and 12 bit. The simulations are performed on 30-pair binders of DTAG-PE06 cable with a line length between 10 and 400 m.

Figure B.28 shows the rate versus reach curve for different linear precoding strategies with co-located CPEs and Fig. B.29 shows the results for non-co-located CPEs.

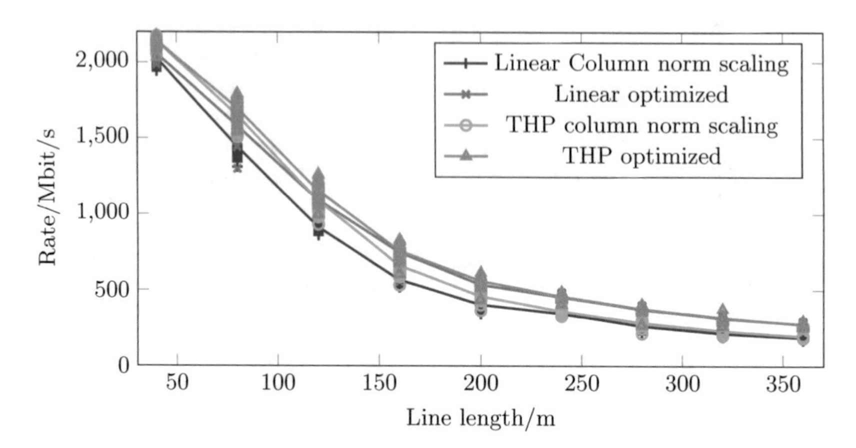

Fig. B.30 Rate versus reach curves for linear and non-linear zero-forcing precoding on a 30 pair DTAG-PE06 binder with co-located CPEs (Fig. 3.23a)

In the more practical non-co-located case, the performance gap between low complexity methods, column norm scaling and greedy sum-rate optimization is much larger than in the co-located case. The performance of spectrum optimization with zero-forcing precoding is close to weighted MMSE precoding, but with significantly lower complexity.

B.18 Simulations for Linear Versus Nonlinear Precoding

The following simulations show the comparison between linear and non-linear (Tomlinson–Harashima) precoding. In both cases, zero-forcing precoding is used with column norm scaling according to Sect. 3.2.6 and with spectrum optimization according to Sect. 3.2.8 for linear precoding and Sect. 3.3.3 for Tomlinson Harashima precoding.

The simulation conditions are a background noise of −140 dBm/Hz, 4 dBm transmit power, and a 212 MHz G.fast profile according to [7] with 2 MHz start frequency. Reed–Solomon and trellis coding with 6 dB target margin is applied, using a discrete bit loading between 1 bit and 12 bit. The simulations are performed on 30-pair binders of DTAG-PE06 cable with a line length between 10 and 400 m.

Figure B.30 gives the results for co-located CPEs and Fig. B.31 the rate versus reach curve for non-co-located CPEs. When comparing linear and non-linear precoding without spectrum optimization, the gain through nonlinear precoding is significant.

Applying spectrum optimization changes the picture, as the spectrum optimized performance of linear and non-linear precoding are very similar. The Tomlinson–Harashima precoding losses, as explained in Sect. 3.3.1 are considered in the simula-

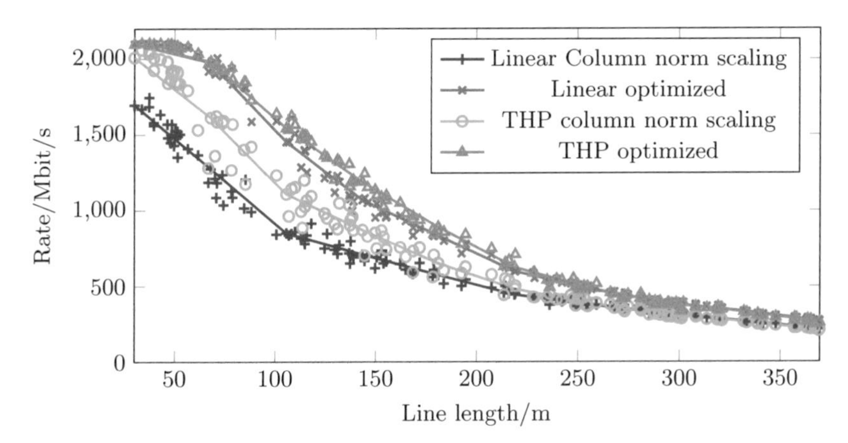

Fig. B.31 Rate versus reach curves for linear and non-linear zero-forcing precoding on a 30 pair DTAG-PE06 binder with non-co-located CPEs (Fig. 3.23b)

tions. On the longer loops, where the THP losses become dominant, linear precoding outperforms nonlinear precoding.

Interference allowing nonlinear precoding is not considered, here. The potential gains of an ideal nonlinear precoding scheme which allows interference and does not experience the losses as they are present for THP is shown in Appendix B.20 as dirty paper coding (DPC).

B.19 Simulations for Upstream and Downstream Performance

Section 3.4 discusses linear equalization for the G.fast upstream.

The simulation conditions are a background noise of −140 dBm/Hz, 4 dBm transmit power, and a 212 MHz G.fast profile according to [7] with 2 MHz start frequency. Reed–Solomon and trellis coding with 6 dB target margin is applied, using a discrete bit loading between 1 bit and 12 bit. The simulations are performed on 30-pair binders of DTAG-PE06 cable with a line length between 10 and 400 m.

Upstream equalization optimization can be solved efficiently, as it is an unconstrained optimization. Figures B.32 and B.33 show the rate vs. reach curves for linear equalization in comparison with the linear downstream precoding rates for zero-forcing with spectrum optimization as well as for the linear MMSE methods.

There is no noticeable difference between MMSE and ZF performance in upstream. Due to the fact that MMSE equalization is easy to implement, it is the preferred method for G.fast upstream equalization.

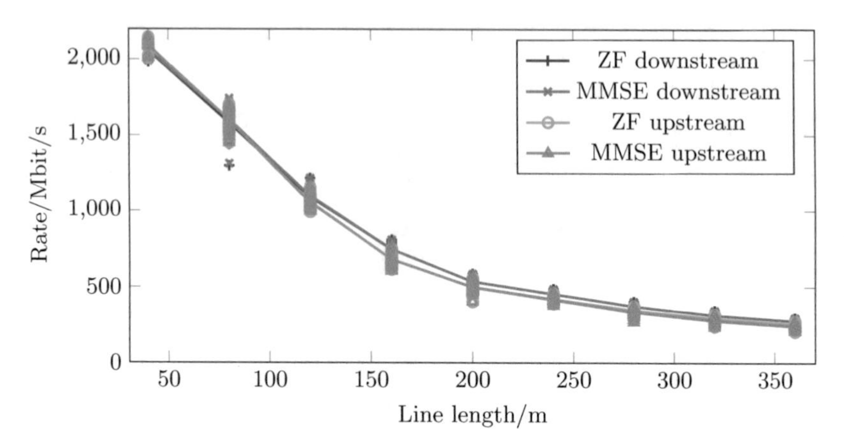

Fig. B.32 Rate versus reach curves for linear downstream precoding and linear upstream equalization on a 30 pair DTAG-PE06 binder with co-located CPEs (Fig. 3.23a)

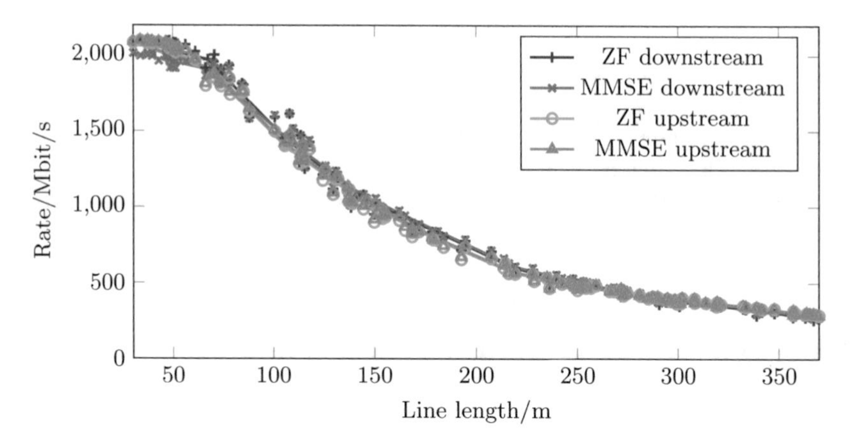

Fig. B.33 Rate versus reach curves for linear downstream precoding and linear upstream equalization on a 30 pair DTAG-PE06 binder with non-co-located CPEs (Fig. 3.23b)

B.20 G.fast Channel Capacity and Loss Analysis

The presented precoding and equalization methods are compared with channel capacity of the G.fast channel, as derived in Sect. 3.5.

The simulation conditions are a background noise of −140 dBm/Hz, 4 dBm transmit power, and a 212 MHz G.fast profile according to [7] with 2 MHz start frequency. The simulations are performed on 30-pair binders of DTAG-PE06 cable with a line length between 10 and 400 m.

Figures B.34 and B.35 show the rate versus reach curves observed with theoretically optimal methods in comparison with the implementable optimized linear zero-forcing and THP schemes.

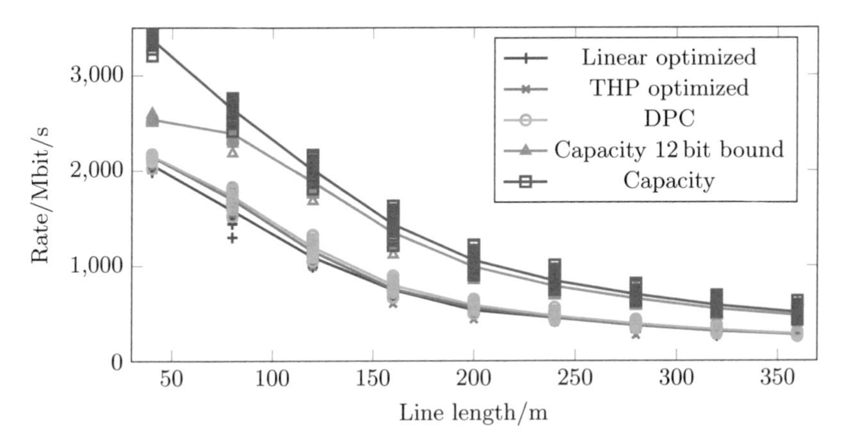

Fig. B.34 Rate versus reach curves for channel capacity and implementation losses on a 30 pair DTAG-PE06 binder with co-located CPEs (Fig. 3.23a)

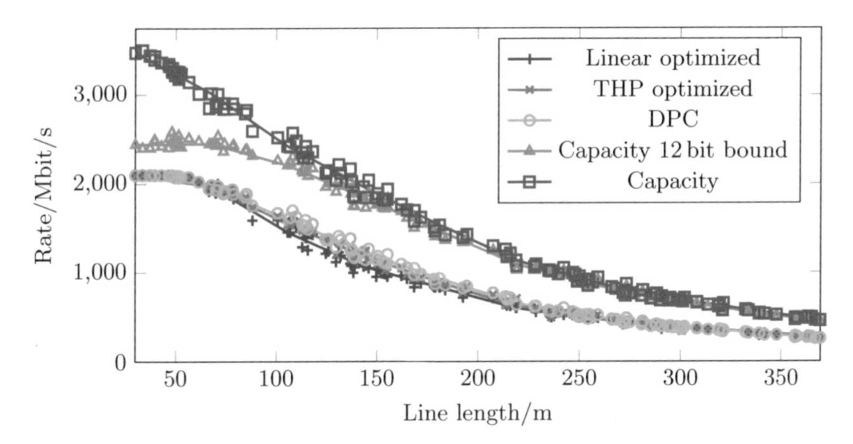

Fig. B.35 Rate versus reach curves for channel capacity and implementation losses on a 30 pair DTAG-PE06 binder with non-co-located CPEs (Fig. 3.23b)

For capacity, fractional bit loading and 0 dB SNR gap are assumed, while for dirty paper coding, Reed–Solomon and trellis coding with 6 dB target margin is applied, using a discrete bit loading between 1 bit and 12 bit, as for implementable precoding methods.

Regarding the unbounded capacity, it must be noted that the transmit spectrum is always limited to 212 MHz while in theory, a wider spectrum may be available on very short twisted pair loops. Therefore, the capacity with infinite spectrum would be higher than shown in Figs. B.34 and B.35, but it is still limited through the line attenuation that increases with frequency.

In the target service reach around 100 m, the capacity drop caused by the 12 bit constraint is not high, which justifies the selection of this design target. The gap between 12 bit-bounded capacity and dirty paper coding (DPC) is significant at any line length for both scenarios. This gap is mainly caused by the imperfections of Reed–Solomon and trellis channel coding and the conservative selection of the target SNR margin of 6 dB.

The simulation results indicate that channel coding still gives some room for performance improvement, while the precoding schemes presented in this work are close to the theoretical optimum achieved by dirty paper coding.

References

1. Hellings, C., Utschick, W.: Performance gains due to improper signals in MIMO broadcast channels with widely linear transceivers. In: International Conference on Acoustics, Speech and Signal Processing (ICASSP), pp. 4379–4383. IEEE (2013)
2. Cioffi, J.: A multicarrier primer. ANSI T1E1 **4**, 91–157 (1991)
3. Zanatta-Filho, D., Lopes, R.R., Ferrari, R., Suyama, R., Dortschy, B.: Bit loading for precoded DSL systems. In: International Conference on Acoustics, Speech and Signal Processing (ICASSP), pp. 353–356 (2007)
4. Le Nir, V., Moonen, M., Verlinden, J., Guenach, M.: Optimal power allocation for downstream xDSL with per-modem total power constraints: broadcast channel optimal spectrum balancing (BC-OSB). IEEE Trans. Signal Process. **57**(2), 690–697 (2009)
5. Guohua, Z., Quan, Z.: Pseudonoise codes constructed by Legendre sequence. Electron. Lett. **38**(8), 376–377 (2002)
6. Strobel, R., Utschick, W.: Coexistence of G.fast and VDSL in FTTdp and FTTC deployments. In: European Signal Processing Conference (EUSIPCO), pp. 1103–1107 (2015)
7. ITU-T Rec. G.9700: Fast access to subscriber terminals (FAST) - power spectral density specification. ITU Recommendation (2013)
8. Maes, J., Nuzman, C.: Energy efficient discontinuous operation in vectored G.fast. In: IEEE International Conference on Communications (ICC), pp. 3854–3858. IEEE (2014)
9. Neckebroek, J., Moeneclaey, M., Coomans, W., Guenach, M., Tsiaflakis, P., Moraes, R.B., Maes, J.: Novel bitloading algorithms for coded G.fast DSL transmission with linear and nonlinear precoding. In: IEEE International Conference on Communications (ICC), pp. 945–951. IEEE (2015)

Appendix C

C.1 Precoder Matrix Update for Discontinuous Operation

The matrix inversion lemma can be used to derive the rule for the precoder coefficient update in discontinuous operation. Starting from the following identity:

$$\begin{bmatrix} \boldsymbol{A} & \boldsymbol{B} \\ \boldsymbol{C} & \boldsymbol{D} \end{bmatrix} \begin{bmatrix} \boldsymbol{E} & \boldsymbol{F} \\ \boldsymbol{G} & \boldsymbol{H} \end{bmatrix} = \boldsymbol{I}, \tag{C.1}$$

which represents, e.g., the product of precoder and channel matrix $\boldsymbol{H}^{(k)}\boldsymbol{P}^{(k)}$, the following identities are found:

$$\boldsymbol{AE} + \boldsymbol{BG} = \boldsymbol{I} \tag{C.2}$$

$$\boldsymbol{AF} + \boldsymbol{BH} = \boldsymbol{0} \tag{C.3}$$

$$\boldsymbol{CE} + \boldsymbol{DG} = \boldsymbol{0} \tag{C.4}$$

$$\boldsymbol{CF} + \boldsymbol{DH} = \boldsymbol{I}. \tag{C.5}$$

Comparing above equations with Eqs. (4.6) and (4.7), the zero-forcing conditions for discontinuous operation, the inverse of the sub-matrix $\boldsymbol{A}$ is needed. Using Eq. (C.2) and inserting Eq. (C.3), the identity

$$\boldsymbol{A}\left(\boldsymbol{E} - \boldsymbol{F}\boldsymbol{H}^{-1}\boldsymbol{G}\right) = \boldsymbol{I} \tag{C.6}$$

is found, which corresponds to the precoder update rule according to Eq. (4.8).

R. Strobel, *Channel Modeling and Physical Layer Optimization in Copper Line Networks*, Signals and Communication Technology,
https://doi.org/10.1007/978-3-319-91560-9

C.2 Precoder Coefficient Update (CU) with Quadratic Complexity

For linear precoding, a special case occurs when limiting the number of discontinued lines per update step L_d to $L_\mathrm{d} = 1$. Hereby, the matrix inversion changes to the inversion of a single scalar. For this case, it is required to select the configurations t such that the set of active lines changes by one in every step, as indicated in Eq. (4.26). Besides that, an additional constraint on the transmit time $\tau^{[t]}$ is required,

$$\tau_t \geq \tau_{\mathrm{update}} \forall t = 1, \ldots, T \tag{C.7}$$

where τ_{update} is the time to compute the new precoder coefficients.

Further setting $\tau_{\mathrm{update}} = \frac{1}{N_{\mathrm{sym}}}$ gives the case that the additional compute complexity for DO coefficient updates is equal to the complexity required for precoding itself.

The update rule for this case is

$$\boldsymbol{P}_{\mathrm{aa}}^{(k),\prime} = \boldsymbol{P}_{\mathrm{aa}}^{(k)} - p_{\mathrm{dd}}^{(k),-1} \boldsymbol{p}_{\mathrm{ad}}^{(k)} \boldsymbol{p}_{\mathrm{da}}^{(k),\mathrm{T}}. \tag{C.8}$$

where $\boldsymbol{p}_{\mathrm{ad}}^{(k)}$ is the the dth column of the precoder matrix, excluding the value $p_{\mathrm{dd}}^{(k)}$ and $\boldsymbol{p}_{\mathrm{da}}^{(k),\mathrm{T}}$ is the dth row of the precoder matrix without the value $p_{\mathrm{dd}}^{(k)}$.

C.3 Signal Flow for Discontinuous Operation Signal Update (SU)

Discontinuous operation for linear precoding can be implemented by changing the precoding operation itself, instead of changing the precoder coefficients to support a changing number of active lines and discontinued lines. Figure C.1 shows the block diagram for the signal processing which are required to perform signal update-based precoding. The input signal vector is $\boldsymbol{u}_\mathrm{a}$ and the transmitted signal vector is $\boldsymbol{y}_\mathrm{a}$ [1].

It is described by

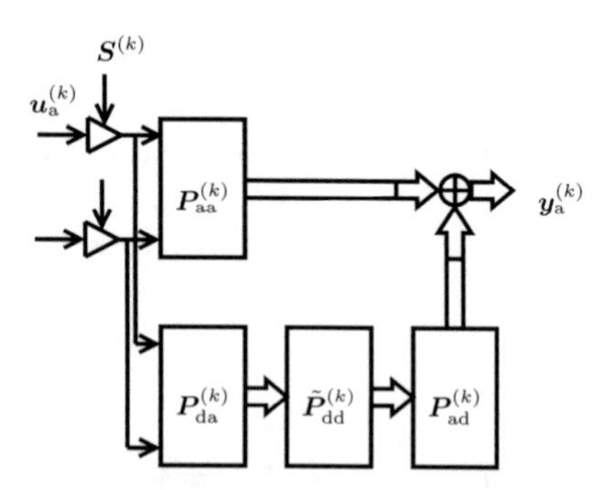

Fig. C.1 Downstream linear precoder supporting discontinuous operation

$$\boldsymbol{y}_{\mathrm{a}}^{(k)} = \boldsymbol{P}_{\mathrm{aa}}^{(k)} \boldsymbol{u}_{\mathrm{a}}^{(k)} + \boldsymbol{P}_{\mathrm{ad}}^{(k)} \tilde{\boldsymbol{P}}_{\mathrm{dd}}^{(k)} \boldsymbol{P}_{\mathrm{da}}^{(k)} \boldsymbol{u}_{\mathrm{a}}^{(k)}. \tag{C.9}$$

with the approximation of the inverse of the matrix $\boldsymbol{P}_{\mathrm{dd}}^{(k)}$ according to $\tilde{\boldsymbol{P}}_{\mathrm{dd}}^{(k)} = \boldsymbol{P}_{\mathrm{dd}}^{(k)} - 2\boldsymbol{I}_{L\mathrm{d}} \approx -\boldsymbol{P}_{\mathrm{dd}}^{(k),-1}$, as described in Sect. 4.3.4.

C.4 Generalized Inverses and Spectrum Optimization

Despite the fact that a G.fast system is assumed to have exactly one DPU transmitter per CPE receiver, i.e., the channel matrix $\boldsymbol{H}^{(k)}$ is always square, there are cases where individual CPE receivers are inactive, resulting in a non-square channel matrix with $L_{\mathrm{a}} < L$ active CPE receivers, such that $\boldsymbol{H}^{(k)} \in \mathbb{C}^{L_{\mathrm{a}} \times L}$.

The situation occurs for certain carriers k when applying spectrum optimization as shown in Chap. 3, because some subcarriers are not used by all lines, e.g., $L_{\mathrm{a}}^{(k)} < L$ due to a low channel quality. Besides that, framing optimization as shown in Chap. 4 may result in a situation where some CPE receivers are not enabled for an individual configuration t, $L_{\mathrm{a}}^{[t]} < L$, and there are times where the precoder matrix is not square. Both cases will result in a non-square precoder matrix for specific carriers k or configurations t.

MMSE precoder and equalizer methods as described in Sects. 3.2.9 and 3.4.2 implicitly support such non-square channel matrices and the algorithms described in the corresponding sections will converge to the desired MMSE optimal solutions.

Zero-forcing methods, based on the constraint $\boldsymbol{G}^{(k)} \boldsymbol{H}^{(k)} \boldsymbol{P}^{(k)} = \boldsymbol{I}$ gain a new degree of freedom in the case of non-square channel matrices, because there is no longer a unique solution for the downstream precoder and the upstream equalizer matrix. In Sect. 3.2.6, it is proposed to compute the non-square precoder based on the pseudo-inverse, e.g.

$$\boldsymbol{P}^{(k)} = \left[\bar{\boldsymbol{H}}^{(k)}\right]^{+} = \bar{\boldsymbol{H}}^{(k)} \left(\bar{\boldsymbol{H}}^{(k)} \bar{\boldsymbol{H}}^{(k),\mathrm{H}}\right)^{-1} \tag{C.10}$$

where $\bar{\boldsymbol{H}}^{(k)} = \boldsymbol{G}^{(k)} \boldsymbol{H}^{(k)}$.

As shown in [2], this is not the only solution for the precoder and when using per-line power constraints, this is not necessarily the optimal solution. Reference [2] introduces the notation of a generalized inverse, where the precoder is given by

$$\boldsymbol{P}^{(k)} = \left[\bar{\boldsymbol{H}}^{(k)}\right]^{+} + \boldsymbol{P}_{\perp}^{(k)} \boldsymbol{U}^{(k)}. \tag{C.11}$$

Hereby, $\boldsymbol{P}_{\perp}^{(k)} = \boldsymbol{I} - \left[\bar{\boldsymbol{H}}^{(k)}\right]^{+} \bar{\boldsymbol{H}}^{(k)}$ is a projection into the Null-space of the non-square channel matrix $\bar{\boldsymbol{H}}^{(k)}$. $\boldsymbol{U}^{(k)}$ is an arbitrary matrix and therefore, the additional degree of freedom for precoder optimization.

The algorithm proposed in [2], which is designed for a per-antenna power constraint, only, cannot be applied for G.fast spectrum optimization, because of the G.fast constraint set consisting of a per-line sum-power and per-line spectral mask constraint.

The spectrum optimization problem can be written as

$$\begin{aligned} \max_{\boldsymbol{x}^{(k)},\boldsymbol{U}^{(k)}} & \sum_{l=1}^{L}\sum_{k=1}^{K} \log_2\left(1+\frac{x_l^{(k)}}{|g_l^{(k)}|^2\sigma^2}\right) \\ & \forall k: \boldsymbol{A}^{(k)}\boldsymbol{x}^{(k)} \leq \boldsymbol{d}^{(k)}\ \forall\, k=1,\ldots,K \\ & \sum_{k=1}^{K} \boldsymbol{A}_{\text{sum}}^{(k)}\boldsymbol{x}^{(k)} \leq \boldsymbol{p}_{\text{sum}}. \end{aligned} \tag{C.12}$$

which gives an additional degree of freedom for zero-forcing spectrum optimization, compared to Eq. (3.61), but still gives a zero-forcing solution for the precoder.

The Lagrange function is given by

$$\begin{aligned} \Phi(\boldsymbol{x}^{(k)}, \boldsymbol{U}^{(k)}, \boldsymbol{\mu}_{\text{sum}}, \boldsymbol{\mu}_{\text{mask}}^{(k)}) &= \sum_{l=1}^{L} -R_l \\ &+ \sum_{k=1}^{K} \operatorname{tr}\left(\operatorname{diag}(\boldsymbol{\mu}_{\text{mask}}^{(k)})\left(\boldsymbol{P}^{(k)}\operatorname{diag}(\boldsymbol{x}^{(k)})\boldsymbol{P}^{(k),\text{H}} - \operatorname{diag}(\boldsymbol{p}_{\text{mask}}^{(k)})\right)\right) \\ &+ \operatorname{tr}\left(\operatorname{diag}(\boldsymbol{\mu}_{\text{sum}})\left(\sum_{k=1}^{K} \boldsymbol{P}^{(k)}\operatorname{diag}(\boldsymbol{x}^{(k)})\boldsymbol{P}^{(k),\text{H}} - \operatorname{diag}(\boldsymbol{p}_{\text{sum}})\right)\right) \end{aligned} \tag{C.13}$$

where the precoder $\boldsymbol{P}^{(k)}$ can be replaced by Eq. (C.11) to have the dependency on $\boldsymbol{U}^{(k)}$.

The dual feasibility conditions are $\frac{\partial\Phi(\boldsymbol{x}^{(k)},\boldsymbol{U}^{(k)},\boldsymbol{\mu}_{\text{sum}},\boldsymbol{\mu}_{\text{mask}}^{(k)})}{\partial\boldsymbol{x}^{(k)}} = \boldsymbol{0}$, which gives the derivative according to Eq. (3.63) and $\frac{\partial\Phi(\boldsymbol{x}^{(k)},\boldsymbol{U}^{(k)},\boldsymbol{\mu}_{\text{sum}},\boldsymbol{\mu}_{\text{mask}}^{(k)})}{\partial\boldsymbol{U}^{(k),*}} = \boldsymbol{0}$, which gives

$$\boldsymbol{P}_{\perp}^{(k),\text{H}}\operatorname{diag}(\boldsymbol{\mu}_{\text{mask}}^{(k)} + \boldsymbol{\mu}_{\text{sum}})\left(\bar{\boldsymbol{H}}^{(k),+} + \boldsymbol{P}_{\perp}^{(k)}\boldsymbol{U}^{(k)}\right)\operatorname{diag}(\boldsymbol{x}^{(k)}) = \boldsymbol{0}. \tag{C.14}$$

The gradient-based method for spectrum optimization according to Sect. 3.2.10 gives a solution for $\boldsymbol{x}^{(k)}$ which satisfies the primal feasibility and complementary slackness conditions as well as the dual feasibility condition with respect to $\boldsymbol{x}^{(k)}$. The matrix $\boldsymbol{U}^{(k)}$ satisfying the dual feasibility condition can be derived with a gradient step

$$\boldsymbol{U}^{(k),[t+1]} = \boldsymbol{U}^{(k),[t]} - \alpha_u \frac{\partial\Phi(\boldsymbol{x}^{(k)}, \boldsymbol{U}^{(k)}, \boldsymbol{\mu}_{\text{sum}}, \boldsymbol{\mu}_{\text{mask}}^{(k)})}{\partial\boldsymbol{U}^{(k),*}} \tag{C.15}$$

with the step size α_u. Alternatively, the lth column $\boldsymbol{u}_l^{(k)}$ of $\boldsymbol{U}^{(k)}$ is given by

$$\boldsymbol{u}_l^{(k)} = \begin{cases} \left(\boldsymbol{P}_\perp^{(k),\mathrm{H}} \mathrm{diag}(\boldsymbol{\mu}_{\mathrm{mask}}^{(k)} + \boldsymbol{\mu}_{\mathrm{sum}}) \boldsymbol{P}_\perp^{(k)}\right)^+ \left[\bar{\boldsymbol{H}}^{(k),+}\right]_l & \text{for } x_l^{(k)} \neq 0 \\ \boldsymbol{0} & \text{for } x_l^{(k)} = 0 \end{cases} \tag{C.16}$$

where $\left[\bar{\boldsymbol{H}}^{(k),+}\right]_l$ is the lth column of the pseudoinverse of the channel matrix.

For non-square precoder matrices, this gives a noticeable performance gain over linear precoding with the Moore–Penrose pseudoinverse, especially for cases where a small number of receiving lines is served by many transmitters.

Still, the method does not achieve the performance of weighted MMSE precoding and adds additional complexity to the precoder and spectrum optimization.

C.5 Power Consumption in G.fast Systems

The most common approach to power minimization in DSL systems, such as [3], aims to minimize the transmit power. Reduction of the transmit power saves energy in the transmit amplifier, the line driver. However, the line driver is only one of many power consumers in the G.fast transceiver. Due to the fact that G.fast technology uses a lower transmit power, usually 4 dBm, than ADSL or VDSL2 which use 14.5 dBm transmit power in most cases, the power saving capabilities in the G.fast line driver are low.

In the G.fast systems, other components than the line driver dominate the over-all power consumption. Government guidelines such as the EU code of conduct [4] define power consumption targets for the complete network equipment rather than for individual components. Version 5 of the code of conduct [4] does not define power consumption targets for G.fast, but the VDSL2 Profile 30 system is allowed to have a power consumption of 1.5 W per port, and the G.fast power consumption target is expected to be in the same range.

It is clear that the G.fast transmit power of 2.5 mW accounts only for a small portion of the power budget. Looking to the block diagram of the G.fast physical layer in Fig. 3.2 in Sect. 3.1, the main building blocks are digital signal processing and analog front-end. The power consumption of these blocks does not scale with transmit power, but they can be switched to low power mode whenever the transceiver does not transmit or receive. This motivates the introduction of discontinuous operation in G.fast.

C.6 Results for Power Saving in Discontinuous Operation

This section gives more details on discontinuous operation simulations for power saving. The simulation results for a DPU size of 16 lines are shown in Figs. 4.7 and 4.8 in Sect. 4.5.1. The simulations use DTAG-PE05 cable binders with 30 lines,

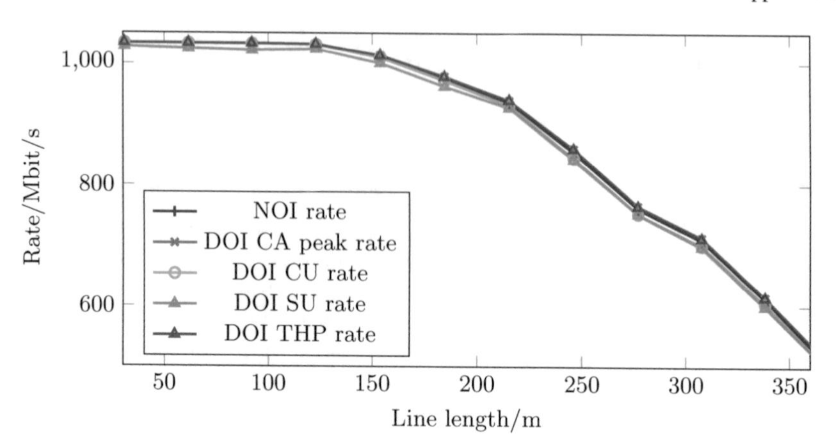

Fig. C.2 Rate versus reach for different discontinuous operation implementations for 4 active lines on a DPU

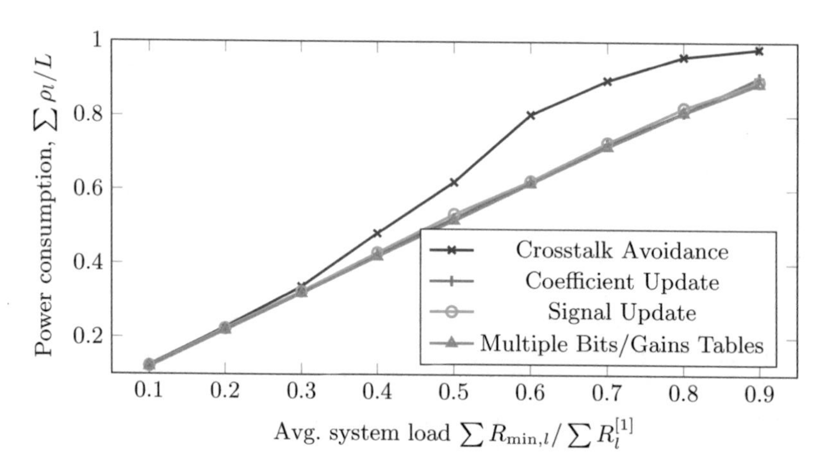

Fig. C.3 Average link on-time over average requested rates for a DPU with 4 lines

where the corresponding number of lines is selected randomly out of the 30 lines. The CPEs are non co-located, e.g., the line length between the DPU and CPE is different for each CPE. The line length is selected randomly between 10 m and 400 m with a uniform distribution. Background noise is assumed to be -140 dBm/Hz below 30 MHz and -150 dBm/Hz above 30 MHz. The transmit power per line is limited to $p_{\text{sum}} = 4$ dBm and the spectral mask according to [5] for a frequency band between 2 MHz and 106 MHz is used.

Linear zero-forcing and zero-forcing Tomlinson–Harashima precoding are used for crosstalk cancelation. Transmit spectrum optimization according to Sects. 3.2.8 and 3.3.3 is applied.

For a DPU size of 4 lines, as shown in Figs. C.2 and C.3, the different DO methods behave very similar and only the crosstalk avoidance method cannot give the full

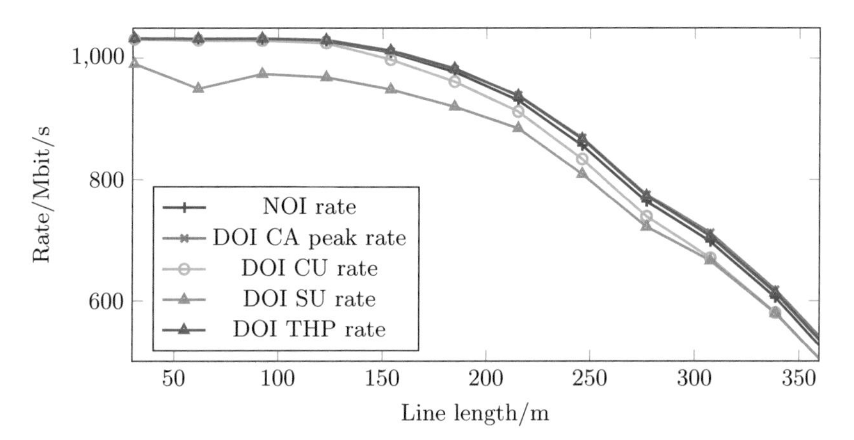

Fig. C.4 Rate versus reach for different discontinuous operation implementations for 8 active lines on a DPU

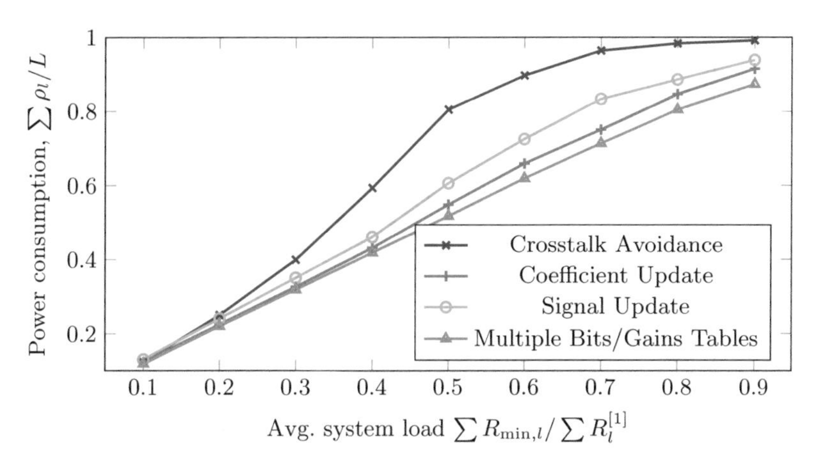

Fig. C.5 Average link on-time over average requested rates for a DPU with 8 lines

power saving at higher loads. The data rates for discontinuous operation and normal operation are very similar.

Increasing the DPU size to 8 lines (Figs. C.4 and C.5) or 12 lines (Figs. C.6 and C.7) increases the gap between the different DO implementations in terms of power saving while the coefficient update method is closest to the optimal solution without implementation limitations. The signal update method requires more transmit time and the crosstalk avoidance scheme requires the highest transmit time at high loads while it is more efficient than the signal update method at lower loads.

The data traffic is modeled by selecting the target data rates $R_{\min,l}$ randomly with an exponential distribution, limit the values between 0 and $0.9R_{\text{NOI},l}$ and scale it such that $\sum R_{\min,l}/\sum R_l^{[1]}$ matches the desired average traffic.

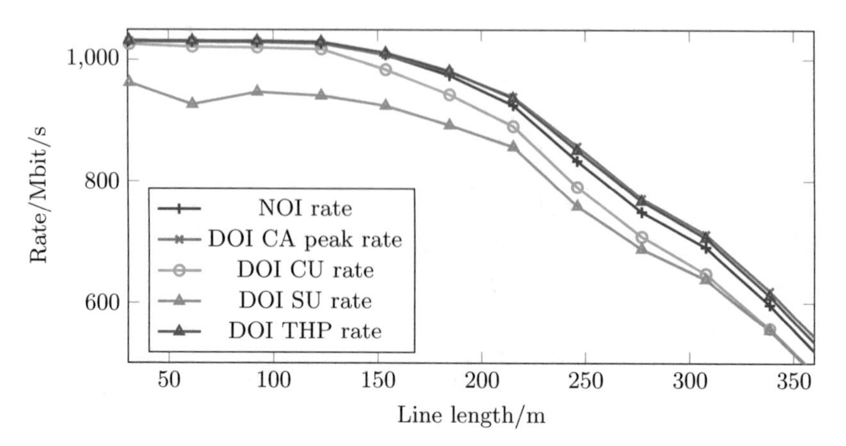

Fig. C.6 Rate versus reach for different discontinuous operation implementations for 12 active lines on a DPU

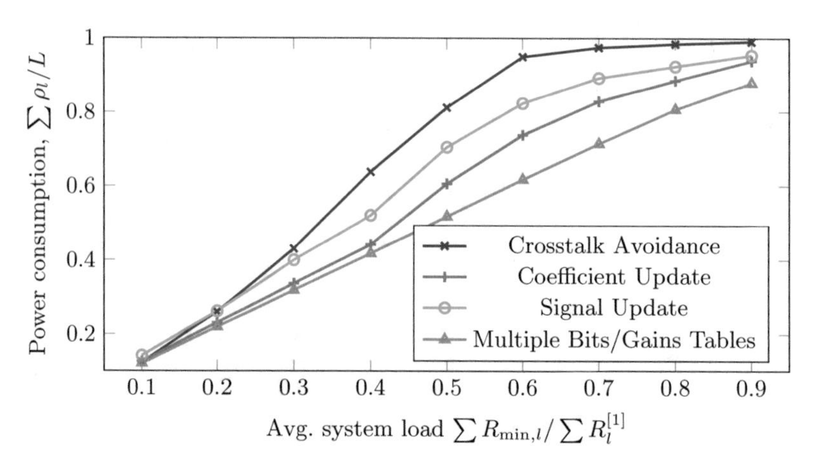

Fig. C.7 Average link on-time over average requested rates for a DPU with 12 lines

C.7 Peak Rate Optimization Simulations

Figures C.8 and C.9 demonstrate the capabilities of peak rate optimization on the DTAG-PE06 cable. Two different network topologies are analyzed. Co-located CPEs according to the network topology shown in Fig. 3.23a in Sect. 3.7 give the simulation results shown in Fig. C.8. A simulation with non-co-located CPEs according to Fig. 3.23b gives the results in Fig. C.9.

The simulations are done on a 30-pair binder with peak rate optimization according to Sect. 4.4 using linear MMSE precoding according to Sect. 3.2.10 for one single line. For comparison, precoded data rates with all 30 lines active are shown for column norm scaling and for linear zero-forcing spectrum optimization on the binder.

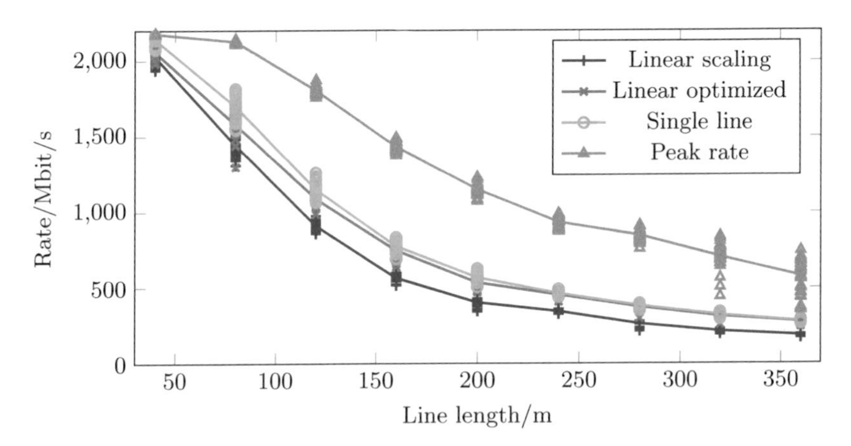

Fig. C.8 Rate versus reach curves for comparison between achievable rates for sum-rate optimization and achievable peak rates on a 30 pair DTAG-PE06 binder with co-located CPEs

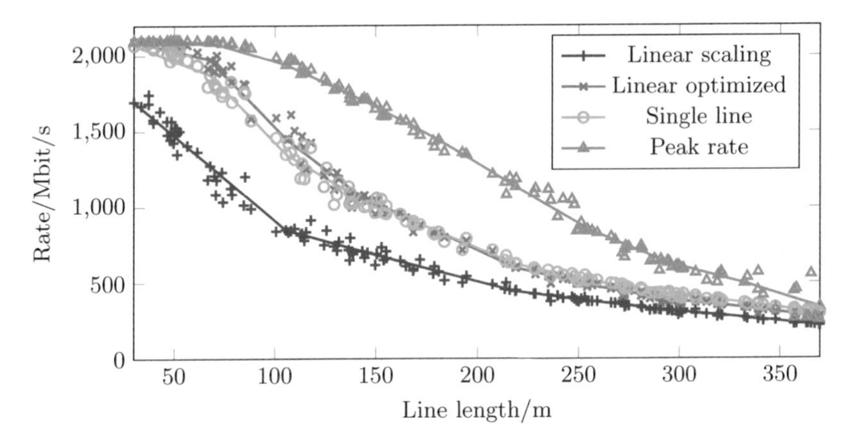

Fig. C.9 Rate versus reach curves for comparison between achievable rates for sum-rate optimization and achievable peak rates on a 30 pair DTAG-PE06 binder with non-co-located CPEs

The single line (SISO) data rates, which are considered for precoding performance comparison in Vectored VDSL2 (e.g, [6]) are used as another reference. For the single line data rates, each line of the binder is enabled independently while only the transmitter and receiver corresponding to the selected line is active. For peak rate optimization, multiple transmitters serve one receiver (MISO).

Figures C.8 and C.9 show that for both network topologies, the achievable peak rates are significantly higher than the achieved rates in the single line case or in the precoded case. Especially in the co-located case, long lines allow more then 100% increase of the peak rates over the data rates of all lines active.

References

1. Huang, Y., Magesacher, T., Medeiros, E., Lu, C., Eriksson, P.E., Odling, P.: Rate-boosting using strong crosstalk in next generation wireline systems. In: IEEE Global Communications Conference (GLOBECOM), pp. 1–6. IEEE (2015)
2. Wiesel, A., Eldar, Y.C., Shamai, S.: Zero-forcing precoding and generalized inverses. IEEE Trans. Signal Process **56**(9), 4409–4418 (2008)
3. Tsiaflakis, P., Yi, Y., Chiang, M., Moonen, M.: Green DSL: energy-efficient DSM. In: IEEE International Conference on Communications (ICC), pp. 1–5. IEEE (2009)
4. European Commission Joint Research Centre: Code of conduct on energy consumption of broadband equipment, Version 5.0. Institute for Energy and Transport (2013)
5. ITU-T Rec. G.9700: Fast access to subscriber terminals (FAST) - power spectral density specification. ITU Recommendation (2013)
6. Cendrillon, R., Ginis, G., Moonen, M.: A near-optimal linear crosstalk precoder for downstream VDSL. IEEE Trans. Commun. **55**(5), 860–863 (2007)

Index

R. Strobel, *Channel Modeling and Physical Layer Optimization in Copper Line Networks*, Signals and Communication Technology,
https://doi.org/10.1007/978-3-319-91560-9

Printed in the United States
By Bookmasters